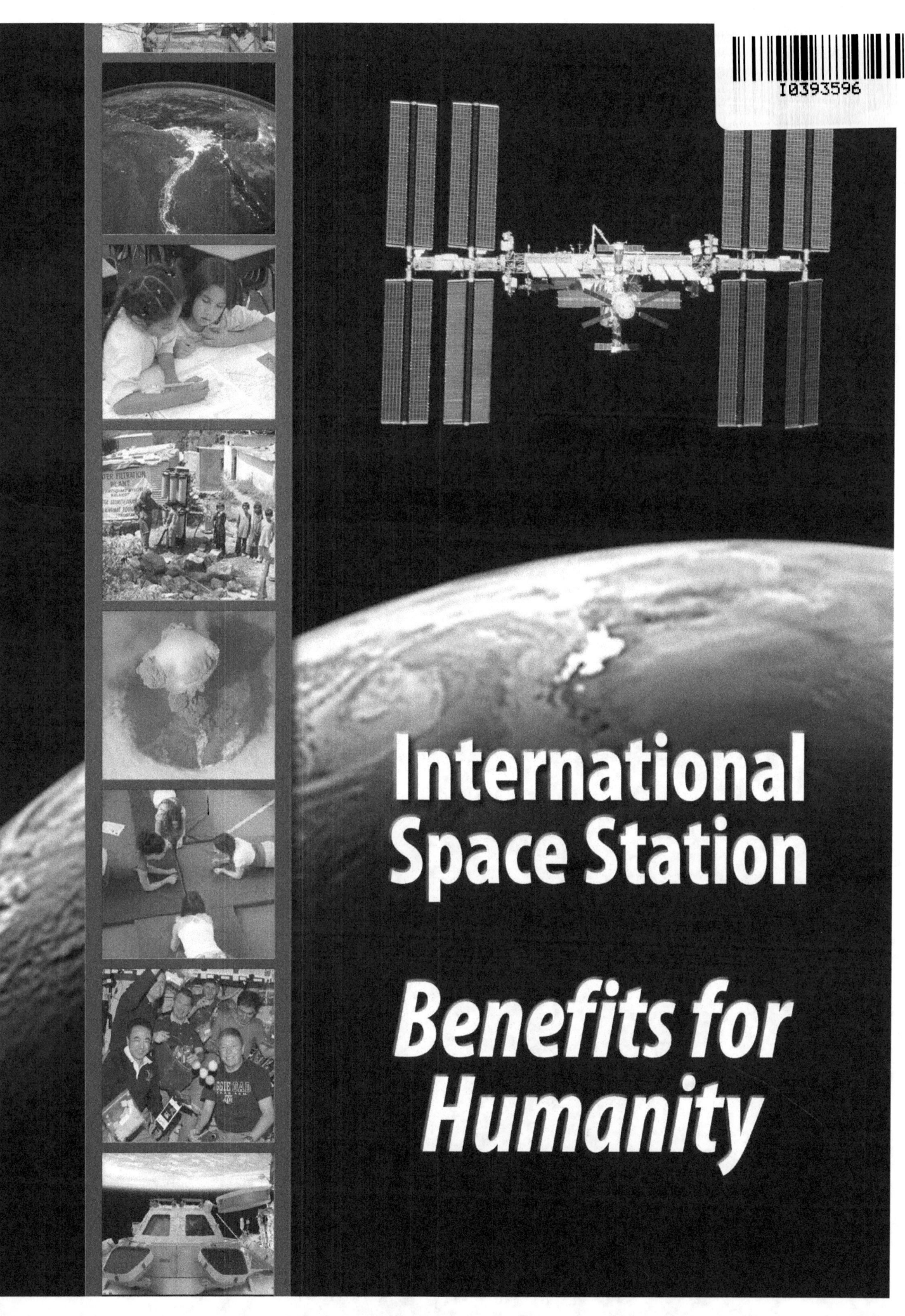

International Space Station

Benefits for Humanity

Related Publications

International Space Station Research and Development Reviews

Comprehensive International Space Station Research Accomplishments

International Space Station Science Research Accomplishments During the Assembly Years: An Analysis of Results from 2000–2008. NASA Technical Paper, 2009; TP-2009-213146-Revision A. *Update expected 2012.*

Benefits of Space Research

International Space Station Benefits for Humanity 2012

Benefits of Space Research, *expected 2012*

International Space Station Education Impacts

International Space Station Benefits for Humanity 2012

Thomas DA, Robinson JA, Tate J, Thumm T. Inspiring the Next Generation: Student Experiments and Educational Activities on the International Space Station, 2000–2006. NASA/TP-2006-213721. 2006; 1-108. *Update expected 2012.*

Other Publications of Interest

Research in Space: Facilities on the International Space Station 2009

Reference Guide to the International Space Station 2010

The Era of International Space Station Utilization: Perspectives on Strategy From International Research Leaders 2010

NP-2012-02-003-JSC

International Space Station Benefits for Humanity

This book was developed collaboratively by the members of the Canadian Space Agency (CSA), European Space Agency (ESA), Japan Aerospace Exploration Agency (JAXA), National Aeronautics and Space Administration (NASA) and the Russian Federal Space Agency (Roscosmos).

The stories contained in this document are a collection of web stories that can be found at www.nasa.gov/mission_pages/station/research/benefits/index.html.

Product of the Program Science Forum

CSA: Nicole Buckley, Perry Johnson-Green, Thomas Piekutowski, Marilyn Steinberg, Jason Clement, Ruth Ann Chicoine

ESA: Martin Zell, Christer Fuglesang, Jason Hatton, Patrik Sundblad, Nigel Savage, Rosita Suenson, Julien Harrod

JAXA: Tai Nakamura, Shigeki Kamigaichi, Ryota Sato, Tatsuya Aiba, Masato Koyama, Yayoi Miyagawa, Shiho Ogawa, Yutaka Kaneko

NASA: Julie Robinson, Tracy Thumm, Tara Ruttley, Camille Alleyne Cynthia Evans, William Stefanov, Elizabeth Richards, Jennifer Fogarty, Judy Carrodeguas, Carolyn Knowles, Regina Blue, Kelly Humphries, Michael Curie, Joshua Buck, Brooke Boen, Kris Rainey

Roscosmos: George Karabadzhak, Igor Sorokin, Sergey Avdeev, Boris Zagreev

Design and Formatting
James Fairchild and Kristi Ferguson

Managing Editor
Tracy L. Thumm, NASA

Executive Editor
Julie A. Robinson, NASA

Contents

Value of the Platform

Engineering Achievement	International Achievement	Research Achievement

Benefits of Research and Technology

Scientific Discovery	On Earth	Space Exploration

Benefits for Humanity Themes

Human Health	Earth Observation and Disaster Response	Education

Executive Summary

Almost as soon as the International Space Station was habitable, researchers began using it to study the impact of microgravity and other space effects on several aspects of our daily lives. This unique scientific platform continues to enable researchers from all over the world to put their talents to work on innovative experiments that could not be done anywhere else. Although each space station partner has distinct agency goals for station research, each partner shares a unified goal to extend the resulting knowledge for the betterment of humanity. We may not know yet what will be the most important discovery gained from the space station, but we already have some amazing breakthroughs! In the areas of human health, telemedicine, education and observations of Earth from space, there are already demonstrated benefits to human life. Vaccine development research, station-generated images that assist with disaster relief and farming, and education programs that inspire future scientists, engineers and space explorers are just some examples of research benefits. This book summarizes the scientific, technological and educational accomplishments of research on the space station that has had and will continue to have an impact on life on Earth.

The benefits outlined here serve as examples of the space station's potential as a groundbreaking scientific research facility. Through advancing the state of scientific knowledge of our planet, looking after our health, and providing a space platform that inspires and educates the science and technology leaders of tomorrow, these benefits will drive the legacy of the space station as its research strengthens economies and enhances the quality of life here on Earth for all people.

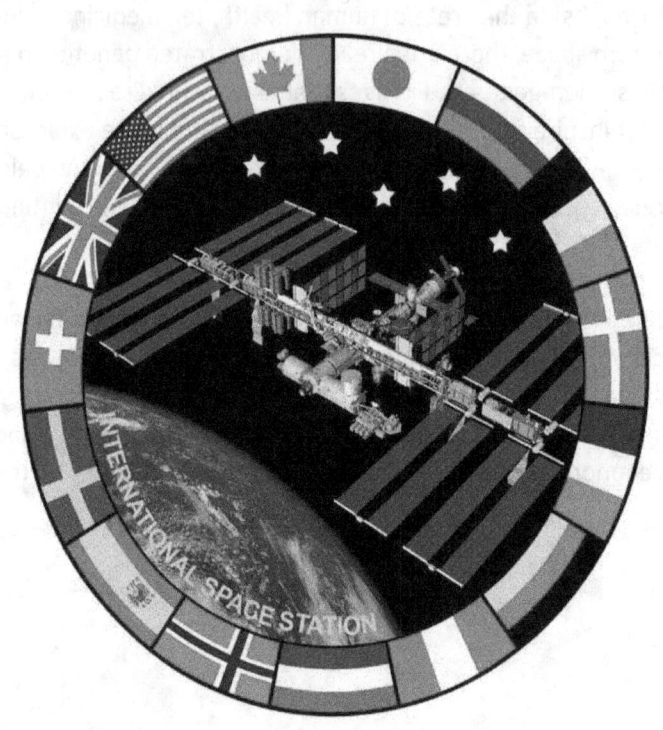

Human Health, Earth Observation, and Education Statement

Human Health

The International Space Station is a unique laboratory for performing investigations that affect human health both in space and on Earth. Throughout its assembly, the space station has supported research that is providing a better understanding of certain aspects of human health, such as aging, trauma, disease and the environment. Several biological and human physiological investigations have yielded important results, including improved understanding of basic physiological processes normally masked by gravity and development of

new medical technology and protocols driven by the need to support astronaut health. Advances in telemedicine, disease models, psychological stress response systems, nutrition, cell behavior and environmental health are just a few examples of benefits that have been gained from the unique space station microgravity environment.

Earth Observation and Disaster Response

The International Space Station is a "global observation and diagnosis station." It promotes international Earth observations aimed at understanding and resolving the environmental issues of our home planet.

The space station offers a unique vantage for observing the Earth's ecosystems with hands-on and automated equipment. These options enable astronauts to observe and explain what they witness in real time. Station crews can observe and collect camera images of events as they unfold and may

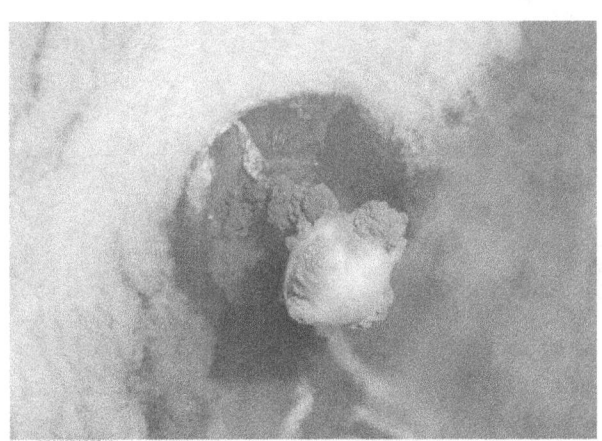

ISS020E009048

also provide input to ground personnel programming the station's automated Earth-sensing systems. This flexibility is an advantage over sensors on unmanned spacecraft, especially when unexpected natural events, such as volcanic eruptions and earthquakes, occur.

A wide variety of Earth-observation payloads can be attached to the exposed facilities on the station's exterior; already, several instruments have been proposed by researchers from the partner countries. The station contributes to humanity by collecting data on the global climate, environmental change and natural hazards using its unique complement of crew-operated and automated Earth-observation payloads.

The existing international partnerships, fundamental to the International Space Station, facilitate data sharing that can benefit people around the world and promote international collaboration on other Earth-observation activities.

Global Education

The space station has a unique ability to capture the imaginations of both students and teachers worldwide. The presence of humans onboard the station provides a foundation for numerous educational activities aimed at capturing that interest and motivating the study of science, technology, engineering and mathematics, or STEM. Projects such as the Amateur Radio on International Space Station, or ARISS; Earth Knowledge Acquired by Middle School Students, or EarthKAM; and Take your Classroom into Space, among others, have allowed for global student, teacher and public

access to space through student image acquisition and radio contacts with crew members. Educational activities are not limited to STEM, but encompass all aspects of the human condition. This is well illustrated in the Uchu Renshi project, a chain poem initiated by an astronaut while in space and continued and completed by people on Earth. With space station operations continuing until at least 2020, projects like these and their accompanying educational materials will be made available to more students around the world. Through the continued use of the station, we will challenge and inspire the next generation of scientists, engineers, writers, artists, politicians and explorers.

neuroArm: Robotic Arms Lend a Healing Touch
Canadian Space Agency

The delicate touch that successfully removed an egg-shaped tumor from Paige Nickason's brain got a helping hand from a world-renowned arm—a robotic arm, that is. The technology that went into developing neuroArm, the world's first robot capable of performing surgery inside magnetic resonance machines, was born of the Canadarm (developed by MDA for the U.S. Space Shuttle Program) as well as Canadarm2 and Dextre, the Canadian Space Agency's family of space robots performing the heavy-lifting and maintenance on board the International Space Station.

neuroArm began with the search for a solution to a surgical dilemma: how to make difficult surgeries easier or impossible surgeries possible. MDA worked with a team led by Dr. Garnette Sutherland at the University of Calgary to develop a highly precise robotic arm that works in conjunction with the advanced imaging capabilities of magnetic resonance imaging (MRI) systems. Surgeons needed to be able to perform surgeries while a patient was inside an MRI machine, which meant designing a robot that was as dexterous as the human hand, but even more precise and tremor-free. Operating inside the MRI also meant it had to be made entirely from nonmagnetic materials (for instance, no steel) so that it would not be affected by the MRI's magnetic field or, conversely, disrupt the MRI's images. The project team developed novel ways to control the robot's movements and give the robot's operator a sense of touch—both essential so that the surgeon can precisely control the robot and can feel what is happening during the surgery.

Since Paige Nickason's surgery in 2008, neuroArm has been used to treat dozens of patients successfully. The neuroArm technology has since been purchased by IMRIS Inc., a private, publicly traded medical device manufacturer based in Winnipeg, Manitoba, Canada. MDA and IMRIS are advancing the design to commercialize a two-armed version of the system to allow surgeons to see detailed, three-dimensional images of the brain as well as use surgical tools and hand controllers that allow the surgeon to feel tissue and apply pressure when he or she operates. A clinical trial led by Dr. Sutherland is currently underway at Calgary's Foothills Hospital using the first generation of the robot on a group of 120 patients. IMRIS anticipates being in a position to seek regulatory approval for the robot as early as 2012.

"Where the robot entered my head," says 21-year-old Paige Nickason, the first patient to have brain surgery performed by a robot, as she points to an area on her forehead. "Now that neuroArm has removed the tumor from my brain, it will go on to help many other people like me around the world."

1

MDA is also continuing to apply its space technologies and know-how to medical solutions for life on Earth. The company has partnered with the Hospital for Sick Children (SickKids) in Toronto, Ontario, to collaborate on the design and development of an advanced technology solution for pediatric surgery. Dubbed KidsArm, the sophisticated, teleoperated surgical system is being designed specifically to operate on small children and babies. KidsArm is intended for use by surgeons in conjunction with a high-precision, real-time imaging technology to reconnect delicate vessels such as veins, arteries or intestines.

In collaboration with The Centre for Surgical Invention and Innovation (CSII) in Hamilton, Ontario, MDA is also developing an advanced platform for use in the early detection and treatment of breast cancer. The image-guided autonomous robot (IGAR) will provide increased access, precision and dexterity, resulting in more accurate and less invasive procedures.

Preventing Bone Loss in Space Flight with Prophylactic Use of Bisphosphonate: Health Promotion of the Elderly by Space Medicine Technologies

Hiroshi Ohshima
Space Biomedical Research Office, JAXA

Bone loss and kidney stones are well known as essential problems for astronauts to overcome during extended stays in space. Crew members engage in physical exercise for two and a half hours a day, six times a week (fifteen hours a week) while in orbit to avoid these issues. Nevertheless, the risks of these problems occurring cannot be completely eliminated through physical exercise alone.

An astronaut performing exercise in the International Space Station. (Photo by JAXA/NASA)

Bone plays an important role as a structure that supports the body and stores calcium. It retains fracture resistance by remodeling through a balance of bone resorption and formation. In a microgravity environment, because of reduced loading stimuli, there is increased bone resorption and no change in or possibly decreased bone formation, leading to bone mass loss at a rate of about ten times that of osteoporosis. The proximal femoral bone loses 1.5 percent of its mass per month, or roughly 10 percent over a six-month stay in space, with the recovery after returning to Earth taking at least three or four years. The calcium balance (the difference between intake and excretion), which is about zero on Earth, decreases to about −250 mg/day during flight, a value that increases the risk of kidney stones.

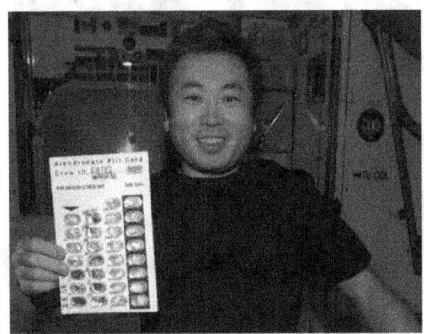

Astronauts take bisphosphonate once a week to prevent bone loss in space. (Photo by JAXA/NASA)

Bisphosphonate is a therapeutic agent that has been used to treat osteoporosis patients for more than a decade, with a proven efficacy to increase bone mass and decrease the occurrence of bone fracture. Through 90-day bed rest research on Earth, we confirmed that this agent has a preventive effect on the loss of bone mass. Based on these results as well as studies conducted by others, JAXA and NASA decided to collaborate on a space biomedical experiment to prevent bone loss during space flight. Dr. Leblanc, USRA, and Dr. Matsumoto, Tokushima University, are the two principal investigators of this study.

JAXA and NASA crew members are participating in this study by taking this agent once a week while in space. Our study is still ongoing; however, early results suggest that astronauts can reduce the risk of bone loss and renal stones by proper intake of appropriate nutrients, such as calcium and vitamin D, an effective exercise program and minimal amounts of medication.

Bone loss is also observed in bedridden older people. Elderly people lose 1 or 2 percent of their bone mass due to aging and a decline in the amount of female hormone. Osteoporosis is declared when a person has a bone mass 30 percent lower than the average for young adults, which is a condition affecting 11 million Japanese and one in two women aged 70 years and older. Every year, 160 thousand patients undergo operations for femoral neck fractures in Japan, followed by intense rehabilitation for three months. Such operations cost 1.5 million yen per person, and the total annual expense for medical treatments and care of these bone fractures amounts to 66.57 billion yen in total national cost.

The three key elements for promoting the health of elderly people to prevent fractures are nutrition, exercise and medicine. Meals should be nutritionally balanced with calcium-rich foods (milk, small fish, etc.) and vitamin D (fish, mushrooms, etc.). Limited sunbathing is also important for activation of vitamin D. Physical exercise to increase bone load and muscle training should also be integrated into each person's daily life. Those at high risk for fractures should take effective medicines to reduce the risk of fractures.

Accordingly, the secrets of the promotion of astronauts' health obtained from space medicine are expected to be utilized to promote the health of elderly people and the education of children.

Astronauts enjoy meals in the International Space Station.
(Photo by JAXA/NASA)

High-Quality Protein Crystal Growth Experiment Onboard "Kibo"

Mika Masaki
Space Environment Utilization Center, JAXA

The vast universe and proteins used to form our bodies are not as unrelated as they may sound—a "crystal" is the key to connect the two. Proteins in the body have three-dimensional, complex structures and the most suitable environment to study these structures is space. Recent studies in Japan, based on the International Space Station research, may help develop more effective methods for treating disease.

In space, there is no convection causing solution to flow up and down and right to left due to differences of density, nor is there precipitation to cause heavier items to sink. Therefore, protein molecules form orderly and create a high quality crystal that is beneficial to the study of its structure. Various crystals have been created in the unique environment of space.

The Japan Aerospace Exploration Agency, or JAXA, has conducted nine sessions of the protein crystallization experiments since 2003 in the Zvezda service module on the International Space Station and has established a technology to produce high-quality protein crystals in space. Based on the technology, a series of six experiments conducted in Kibo began in July 2009, and completion is expected, by the beginning of 2013.

With the support of Russian Federal Space Agency known as Roscosmos, the protein samples installed in the Cell Units (Fig. 1) are launched to the space station on board the Russian Progress spacecraft. Soon after the docking, the samples are brought into Kibo to be placed inside the Protein Crystallization Research Facility, or PCRF, where they will be kept for a period of two to four months at a stable temperature, 68 degrees Fahrenheit (20 degrees Celsius) (Fig. 2). A counter-diffusion method called "Gel-Tube method," is used for crystallization whereby polyethylene glycol or salt solution is diffused into the protein solution separated by a porous membrane inside a tube. Then, concentration of polyethylene glycol in the protein solution gradually increases and finally satisfies the condition for protein crystallization (Fig. 3).

Figure 1. Cell Unit

Cell Unit is used for incubation of the proteins when it is installed into the Protein Crystallization Research Facility or PCRF on board the space station. It is also a transportation container of the Gel-Tube cartridges as it is loaded into the cargo vehicles.

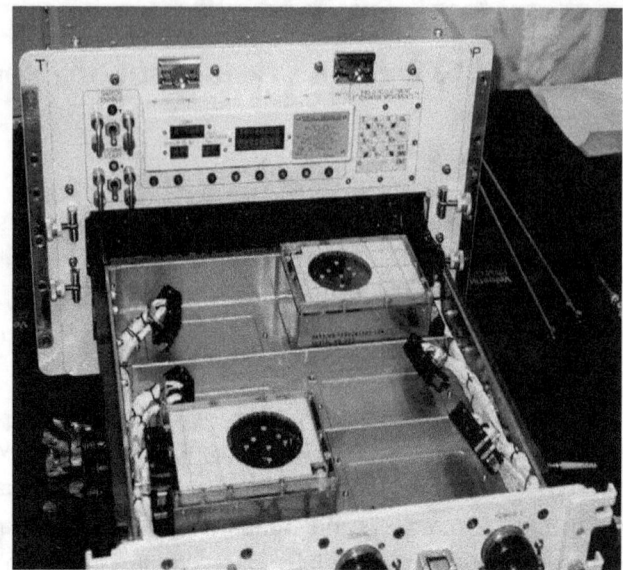

Figure 2. Protein Crystallization Research Facility

Protein Crystallization Research Facility, or PCRF, is a sub-rack payload used for the protein crystallization experiments, which provides controlled temperature and can hold six cell units (up to 144 proteins) inside.

Figure 3. Gel Tube Cartridge (JCB)

Gel-Tube cartridge called "JCB" which contains twelve crystallization capillaries inside is a simple and economical crystallization tool. It requires only minimum onboard operation of astronauts. (© JAXA)

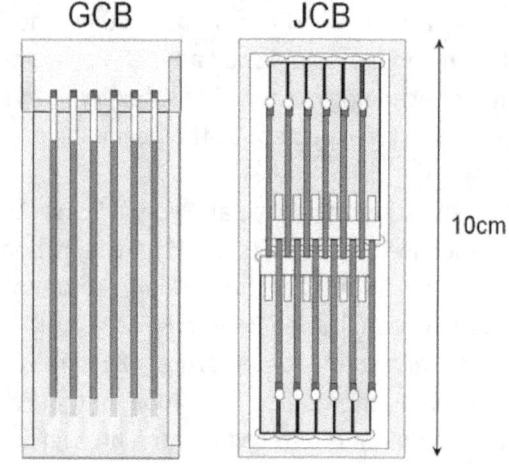

A major purpose of the protein crystal growth experiment is its contribution to the development of medical treatments. The protein that causes disease and the medicine that suppresses it can be compared to the relation between a "keyhole" and a "key." If the shape of the keyhole is discovered by examining the structure of the protein, treatment-oriented medicine with few side effects—the key to fit the keyhole—can be designed. JAXA is making positive advancements in their research on obstinate diseases through experiments in space with the hope of supporting medical care more effectively (Fig. 4).

An example of a protein that was successfully crystallized in space that may hold the key to treating a disease is hematopoietic prostaglandin D synthase or H-PGDS. H-PGDS is the enzyme responsible for the production of hematopoietic prostaglandin D synthase 2 or PGD2, which is known to be a mediator of allergic and inflammatory reactions. Recently, a research team at the Osaka Bioscience

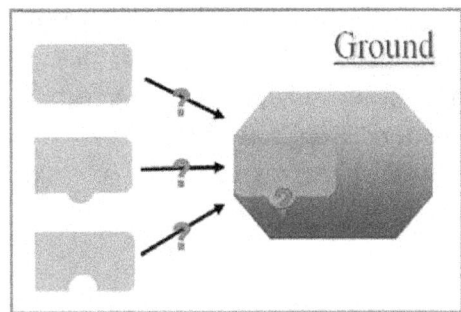

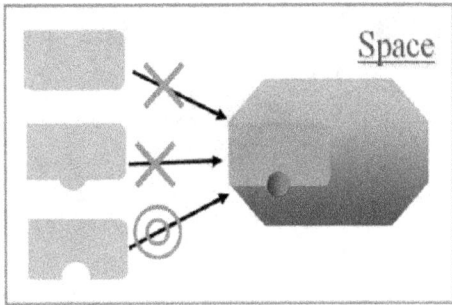

Figure 4. Advantage of the Space Experiment

Because the structure of the disease-causing protein, or the keyhole, is vague when it is obtained on the ground, the shape of the key, or a medicine candidate compound for treatment, cannot be determined. However, it is possible to find the structure of the disease-causing protein through the space experiments, and medicine that fits the treatment (the key that fits the keyhole) can be developed. (© JAXA)

Institute or OBI reported that H-PGDS is expressed in necrotic muscle fibers of patients with Duchenne muscular dystrophy or DMD. An inherited muscle disorder, DMD is the most common form of muscular dystrophy, affecting approximately 1 in 3,500 boys. DMD causes muscular atrophy and accelerates the progression of muscular deterioration. It is an obstinate disease for which a fundamental mode of treatment has not yet been found. Therefore, H-PGDS-specific inhibitors are considered to be useful drugs for muscular dystrophy.

The OBI research team has successfully determined the three-dimensional structure of H-PGDS in a complex with a prototype H-PGDS-specific inhibitor. H-PGDS has been crystallized several times in microgravity as part of JAXA's space experiments. Using X-ray crystallographic analysis—using X-rays to determine the structure—researchers determined the structure of the high-quality crystals of H-PGDS-inhibitor complexes diffracting to 1.0–1.5 Angstrom resolution in space, and discovered a new inhibitor with several hundred times stronger activity than the prototype inhibitor (Fig. 5). This provides a better understanding of the binding mode of H-PGDS with inhibitors for better drug design. Further research is ongoing with an aim of practical use (Fig. 6).

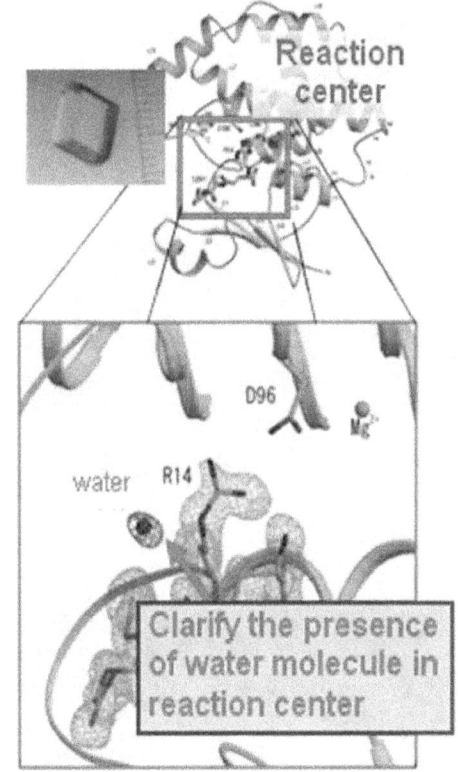

Figure 5. High-Quality Crystals of H-PGDS-Inhibitor Complexes
The detailed structure of muscular dystrophy related protein became clear through a space experiment. (© Osaka Bioscience Institute/MARUWA Foods and Biosciences, Inc.)

There are more than 100,000 proteins in the human body and as many as 10 billion in nature. Every structure is different and each one of them holds important information related to our health and to the global environment. Space is now being utilized as the latest site of biomedical research, where scientists conduct experiments that are impossible or extremely difficult on Earth to understand the structures of proteins. Various investigations in Kibo generate high-quality protein crystals open doors to new possibilities. The elucidation of protein structure is instrumental to understanding the mechanisms of life.

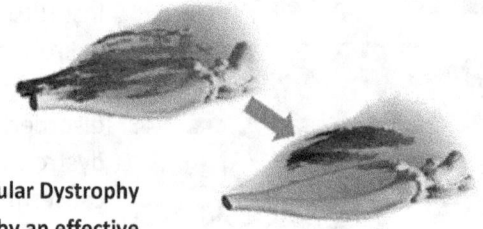

Figure 6. Treatment for Muscular Dystrophy
The progress of muscular atrophy is reduced by an effective medicine candidate compound. (© Osaka Bioscience Institute)

Are You Asthmatic? Your New Helper Comes From Space.
European Space Agency

Kalle, a 10-year-old boy, is already in favor of space technology. In the future, he could control his asthma with a small device also used by crew members on board the International Space Station. Because of it, he knows almost everything about nitric oxide—an important gas we all breathe out.

Nitric oxide, or nitrogen monoxide, as it is properly called, is both a good and bad molecule, found almost everywhere as an air pollutant that is produced by vehicle exhaust and industrial processes burning fuel. Nitric oxide is a contributor to the damage of the ozone layer and easily converts into nitric acid—which may fall as acid rain.

Intriguingly, tiny amounts of nitric oxide are released locally in inflamed tissue of humans and other mammals. Tracing it back to its source can reveal different diseases.

In people with asthma, inflammation in the lung adds nitric oxide to exhaled air. Measuring the gas can help to diagnose the disease and may prevent attacks if the levels of nitric oxide indicate that medication should be adjusted.

Nitric oxide is also an interesting molecule on the space station. Dust and small particles floating around in weightlessness can be inhaled by the astronauts, possibly triggering inflammation of the airways. It also plays a role in decompression sickness that may arise from spacewalks.

The European Space Agency, or ESA, uses a lightweight, easy-to-use, accurate device for measuring nitric oxide in exhaled air. The aim is to investigate possible airway inflammation in astronauts and act before it becomes a health problem.

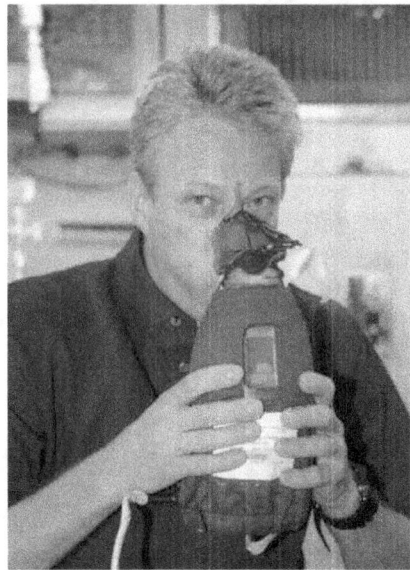

Former ESA astronaut Thomas Reiter undertaking science activities for the Nitric Oxide Analyzer (NOA) experiment in 2006. (Image: ESA)

Following its development by the Swedish company Aerocrine AB and ESA, the device has been found beneficial in space exploration and everyday use on Earth.

NIOX MINO® is now used by patients like Kalle at health centers. They can monitor levels of asthma control and the efficiency of medication—leading to more accurate dosing, reduced attacks and improved quality of life.

For further information, please contact:
Lars Karlsson and Lars Gustafsson
Karoliniska Institutet, Stockholm, Sweden
Dept of Physiology and Pharmacology
Karolinska Institutet
Nanna Svartz väg 2
S-171 77 Stockholm, Sweden
Tel. +46 8 524 868 90
Email: Lars.Karlsson@ki.se

Add Salt? Astronauts' Bones Say Please Don't.

European Space Agency

Osteoporosis is a harsh disease that reduces the quality of life for millions and costs Europe around €25 billion ($31 billion) each year. It typically affects the elderly, so the rise in life expectancy in developed countries means the problems inflicted by osteoporosis are increasing.

Fortunately, research done in space may change the game. Astronauts on the International Space Station experience accelerated osteoporosis because of weightlessness, but it is carefully controlled, and they can regain their lost bone mass once they are back on Earth.

Studying what happens during long spaceflights offers a good insight into the process of osteoporosis—losing calcium and changing bone structure—and helps to develop methods to combat it.

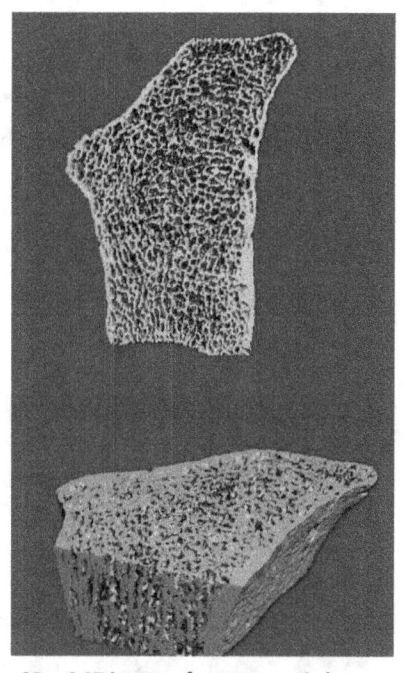

3D pQCT image of osteoporotic bone. (Image: Scanco Medical AG)

It has been known since the 1990s that the human body holds on to sodium, without the corresponding water retention, during long stays in space. But the textbooks said this was not possible. "Sodium retention in space" became an important subject to study.

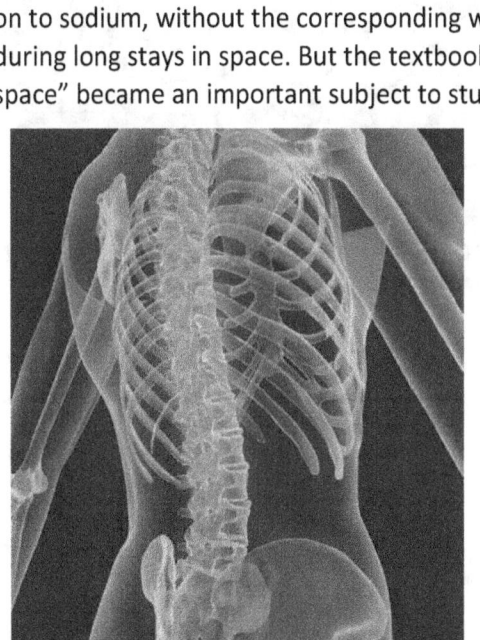

Salt intake was investigated in a series of studies, in ground-based simulations and in space, and it was found that not only is sodium retained (probably in the skin), but it also affects the acid balance of the body and bone metabolism. So, high salt intake increases acidity in the body, which can accelerate bone loss.

The European Space Agency's, or ESA's, recent SOdium LOad in microgravity, or SOLO, study zoomed in on this question.

The SOLO experiment is carrying out research into salt retention and its effect on bone metabolism in astronauts, which can help provide insights into medical conditions on Earth, such as osteoporosis. (Image: Istockphoto/S.Kaulitzki)

Nine crew members, including ESA's Frank De Winne and Paolo Nespoli during their long-duration flights in 2010 and 2011, followed low- and high-salt diets. The expected results may show that additional negative effects can be avoided either by reducing sodium intake or by using a simple alkalizing agent like bicarbonate to counter the acid imbalance.

This space research directly benefits everybody on Earth who is prone to osteoporosis.

For further information, please contact:

Petra Frings-Meuthen
German Aerospace Center (DLR)
Institute of Aerospace Medicine
Space Physiology
Linder Höhe
D-51147 Cologne, Germany
Tel: +49 2203 601-3034
Fax: +49 2203 61159
Email: petra.frings-meuthen@dlr.de
Web: http://www.dlr.de

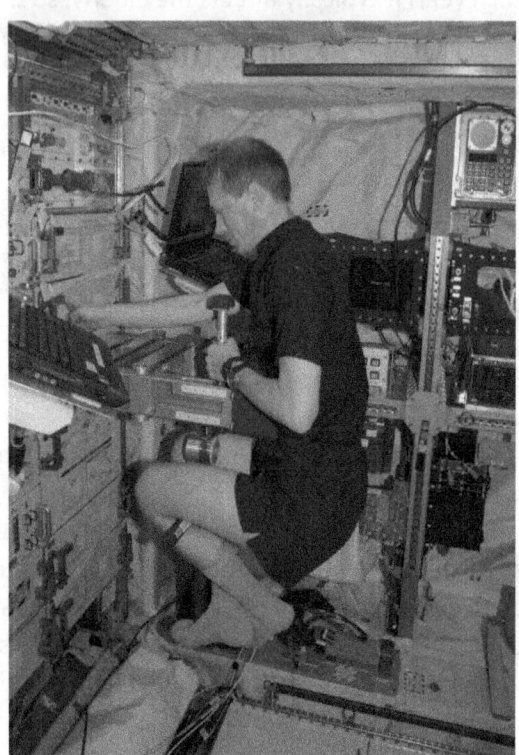

ESA astronaut Frank De Winne undertaking a body mass measurement, an essential element of the SOLO experiment, on the space station.
(Image: ESA)

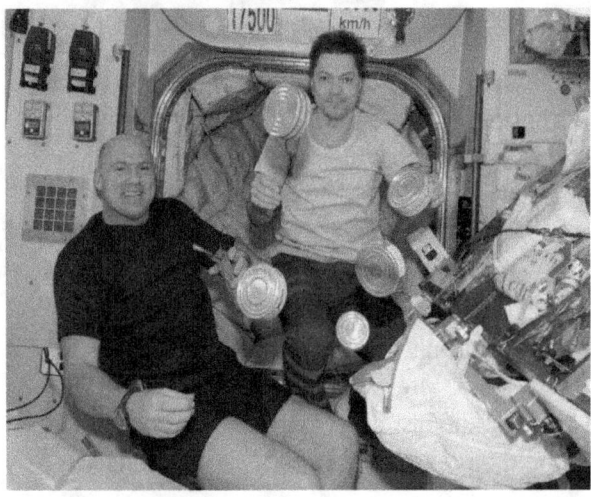

ESA astronaut André Kuipers (left) and Russian cosmonaut Oleg Kononenko (right) with food items on the ISS in December 2011. In the SOLO experiment, astronaut subjects undergo two different diet regimes to determine the physiological effects of sodium on the body.
(Image: ESA)

International Space Station Plays Role in Vaccine Development

Tara Ruttley, Ph.D., Associate International Space Station Program Scientist
NASA

Have you ever been afflicted with a case of food poisoning so awful it made you stop to wonder why no one's found a cure or sure-fire preventative for it yet? And chances are you or someone you know has experienced a bacterial staph infection so aggressive it was resistant to nearly every antibiotic used by the medical profession. The development of vaccines to different pathogens has impacted our global health in ways we could have never anticipated as recently as the early 20th century, and there are still plenty of pathogens to protect ourselves from. The evolution of vaccine development is being streamlined with the help of the microgravity environment exhibited on the International Space Station.

Researchers Timothy Hammond, Ph.D., at the Durham Veterans Affairs Medical Center and Cheryl Nickerson, Ph.D., at Arizona State University have both flown experiments, using microgravity in their search for therapeutic agents or vaccines against Salmonella bacteria. Salmonella infection is one of the most common forms of food poisoning in the U.S. Worldwide, Salmonella diarrhea remains one of the top three causes of infant mortality, so a vaccine has the potential to make dramatic improvements in health for developing countries. The space environment has been shown to induce key changes in microbial cells that are directly relevant to infectious disease, including alterations of microbial growth rates, antibiotic resistance, microbial invasion of host tissue, organism virulence (the relative ability of a microbe to cause disease) and genetic changes within the microbe. Collectively, this body of work has shown that the virulence of this organism increases in microgravity. The targets identified from each of these microgravity-induced alterations represent an opportunity to develop new and improved therapeutics, including vaccines, as well as biological and pharmaceutical agents aimed specifically at eradicating the pathogen.

Early work that laid the foundation for the microgravity-based vaccine development studies began in 1998, when Nickerson initially was funded by NASA in an effort to understand how Salmonella bacteria would respond to a microgravity environment. This was the first of what would be multiple studies from this team on Salmonella bacteria grown in true microgravity or ground-based analogues of microgravity.

Cheryl Nickerson of the Biodesign Institute at Arizona State University (Credit: Nick Meek)

Astronaut John Phillips, STS-119 mission specialist, activates the MSRA experiment on the middeck of Space Shuttle Discovery. (Image credit: NASA)

Follow-on experiments conducted on space shuttle flights to the space station have examined the virulence of methicillin-resistant Staphylococcus aureus, known as MRSA, as well as other microbes. MRSA is a type of Staphylococcus bacteria that is resistant to certain beta-lactam antibiotics; these antibiotics include methicillin, penicillin, and amoxicillin. More severe or potentially life-threatening MRSA infections occur most frequently among patients in healthcare settings. MRSA is especially troublesome in hospitals, where patients with open wounds, invasive devices and weakened immune systems are at greater risk of infection than the general public.

The studies of Salmonella and MRSA bacteria in space are part of the U.S. National Laboratory pathfinder program to demonstrate the use of the space station as a research platform for commercial research and development. The pathfinder research approach uses a set of flight experiments to identify the components of the organisms that facilitate increased virulence in space, and then applies that information to pinpoint targets for anti-microbial therapeutics, including vaccines. Discovering the factors responsible for growth and virulence of bacteria will contribute to the development of novel therapeutic treatments, including vaccines. In fact, the commercial corporation Astrogenetix's space-based Salmonella research has resulted in the discovery of a potential candidate vaccine for this pathogen and is currently in the planning stages for review and commercial development.

More recently, two Arizona State University teams led by Nickerson and Roy Curtiss III, Ph.D., worked together to deliver vaccine samples that were flown to the space station on STS-135. The investigation seeks to improve on existing vaccines against Streptococcus pneumonia—a bacteria that causes life-threatening diseases, such as pneumonia, meningitis, and bacteremia. This organism is responsible for more than 10 million deaths annually and is particularly dangerous for newborns and the elderly, as they are less responsive to current anti-pneumococcal vaccines

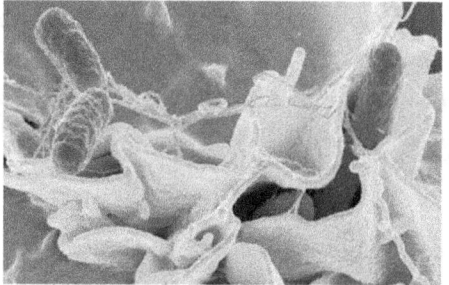

An example of Salmonella invading cultured human cells. (Image courtesy of Rocky Mountain Laboratories, NIAID, NIH)

traditionally delivered by needle. An orally delivered vaccine known as the Recombinant Attenuated Salmonella Vaccine, or RASV, is currently undergoing clinical trials, and the Arizona State University research teams are seeking to increase this vaccine's anti-pneumococcal effectiveness by maximizing its ability to induce a protective immune response. "We have the opportunity," commented Nickerson, "to utilize spaceflight as a unique research and development platform for novel applications with potential to help fight a globally devastating disease." The samples sent to the space station were a genetically altered strain of Salmonella that carries a protective antigen against the Streptococcus pneumonia bacteria. Molecular targets identified from this work hold promise for translation to develop new and improve existing anti-pneumococcal RASVs to prevent disease for the general public. Moreover, because RASVs can be produced against a wide variety of human pathogens, the outcome of this study could influence the development of vaccines against many other diseases in addition to pneumonia.

This space-based research provides evidence that the International Space Station as a National Laboratory is a valuable resource that can be utilized for the benefit of Earth. Discovery of therapeutic targets for MRSA and Salmonella infections are examples of efforts to use the novel microgravity environment to develop new pharmaceutical agents, and as the station nears completion, there will be an increase in such opportunities to utilize the International Space Station National Laboratory as a platform for drug discovery. Overall, these results represent just a fraction of the possibilities of future microgravity discoveries. Scientists participating in these studies plan to fly a continuing series of experiments to the space station, giving them streamlined access that will help to accelerate progress for several different lifesaving vaccines.

Advanced Ultrasound for the Space Program and on Earth[1]

Scott A. Dulchavsky, M.D.[2]
Henry Ford Hospital

Kathleen Garcia; Douglas R. Hamilton, M.D., Ph.D.; Shannon Melton; and Ashot E. Sargsyan, M.D.[3]
Wyle Integrated Science and Engineering

A patient's location at the time of a medical crisis often determines their pain and suffering and even their chances of survival. As a rule, ease of access to medical care decreases as the distance from a developed metropolitan area increases. Providing medical care for people in remote communities; at research outposts, such as Antarctic stations; and on isolated crews, such as the International Space Station crew, is particularly challenging. Medical care at these remote locations is usually performed by minimally trained medical personnel, and a physician is sometimes available only through phone or Internet links, if at all. The ability to quickly diagnose an illness or injury and initiate treatment improves the outcome for the patient and reduces the consequences for the rest of the mission. The ability to make an accurate diagnosis in remote areas reduces the impact of the incident and the chances of an expensive and potentially dangerous and unnecessary evacuation.

Ultrasound imaging is among the fastest, safest and most universal diagnostic methods ever invented. It provides much of the information that can be obtained by expensive technologies, such as X-ray, computed tomography, or magnetic resonance imagery, and it is the only method to produce a real-time or live image that can be interpreted and/or transmitted at the same time. In the right hands, ultrasound instantly answers many clinical questions, shortening the assessment time and improving outcome.

The NASA Advanced Diagnostic Ultrasound in Microgravity, or ADUM, research team, based at NASA's Johnson Space Center, tests novel uses of ultrasound in large medical centers and laboratory conditions, and then adapts them for use in space flight, providing training and guidance to non-medical operators. The ADUM team is comprised of individuals with unique clinical, scientific, engineering, and teaching expertise, as well as direct experience in a multitude of telemedicine projects and programs.

Earth-Based Testing: Expanded Uses for Ultrasound

Ultrasound is routinely used to obtain diagnostic information about pregnancy and various abdominal and vascular conditions such as gallbladder disease or blood clots. We have examined

1 Adapted from an original article that appeared in NASA Technology Innovation, Vol 15; 3, 2010; NP-2010-06-658-HQ.
2 Scott Dulchavsky, M.D., is chair of surgery at Henry Ford Hospital in Detroit, Mich., and the lead investigator in the ADUM project.
3 Kathleen Garcia; Douglas Hamilton, M.D., Ph.D.; Shannon Melton; and Ashot Sargsyan, M.D., are with Wyle Integrated Science and Engineering and are co-investigators in the ADUM project. Wyle is the prime contractor for the NASA Johnson Space Center Bioastronautics contract, providing medical operations, ground and flight research, space flight hardware development and fabrication, science and mission integration.

the use of ultrasound in additional conditions, including collapsed lung, broken bones, injuries to the eyes or head and infections in the teeth or sinus cavities.

Patients who have had chest injuries are at risk of a collapsed lung or pneumothorax, which is usually diagnosed with a chest X-ray. The ADUM team developed a simple lung ultrasound technique that can be used to diagnose a pneumothorax with higher accuracy than a chest X-ray. This new technique is now standard in many hospitals and trauma centers around the world. Using parabolic flight as an analog to microgravity, the ultrasound hardware, astronaut training and remote guidance procedures were validated for use on the space station as a potential capability for the medical support of future space missions.

The ADUM team developed "cue cards" to rapidly guide nonexpert users to perform ultrasound examinations on patients with extremity injuries and found that broken bones can be diagnosed with more than 90-percent accuracy after just minutes of training. Ultrasound also can be used to determine if muscles, joints or tendons are injured. Importantly, it provides a convenient way to look at muscles and joints as they move, which is an advantage over X-ray or other techniques that only provide still images of the body.

There is concern that astronauts will suffer eye injuries caused by objects floating in the spacecraft and, more recently, the effect of prolonged exposure to microgravity on vision. Ultrasound can be used to determine if there are foreign particles in the eye and to identify other conditions that could affect eyesight during space travel. Ocular ultrasound also can provide important information about the condition of the brain in head-injured patients. Those with brain swelling can be identified by using ultrasound to measure the size of the nerve in the back of the eye.

Rapid Training for Nonexpert Operators

During an ultrasound examination, a probe is placed on the patient's body to transmit and receive sound waves to produce a moving image. The technique depends on proper placement and movement of the probe to obtain the best images, and it generally requires hundreds of hours of practice. The ADUM investigators found that it is possible to use non-medical operators to obtain good-quality data if the right clinical questions are asked and the operator is given the right amount and type of information and direction from a remotely located expert.

The ADUM team developed a bilingual (English and Russian) computer-based On-Board Proficiency Enhancement, or OPE, e-learning tool, which consists of a stepwise program for performing targeted ultrasound examinations after only a short hands-on training program before flight. The OPE program includes modules that review equipment set up, basic and advanced ultrasound principles, anatomy, remote guidance principles and exam-specific suggestions with a reference collection of target images. Companion cue cards, which show where to apply the probe to obtain the correct image, were developed for all of the ultrasound examination sets. This training regimen can be completed in two or three hours and includes 30-minute refresher modules to be completed just before an examination.

The key concept developed by the ADUM team is the methodology for "Remote Expert Guidance" of ultrasound examinations, which links a remote expert with the on-site operator in a virtual common working environment. The ultrasound machine video output is transmitted to the remote expert via a satellite or Internet connection, and the operator is guided to obtain the ultrasound images via voice commands. This technique has dramatically reduced training requirements (often down to minutes) while preserving the quality of the ultrasound examinations.

Ultrasound on the Space Station

Initial trials of the new paradigm, using ultrasound to capture images of the heart and abdomen, led the way to a large series of ADUM experiments on the station, sponsored by NASA and the National Space Biomedical Research Institute, or NSBRI. The ADUM investigators stressed the importance of minimal preflight training of astronaut and cosmonaut crew members, computer-based refresher e-learning before imaging sessions, and complex ultrasound examinations on the station with remote expert guidance.

Crew members first performed diagnostic quality cardiac, vascular and thoracic ultrasound examinations on the station. Then other astronauts were rapidly trained to expand the capabilities to perform examinations of the heart, lungs, blood vessels and abdomen with a special emphasis on musculoskeletal ultrasound. The crew members performed bone ultrasound examinations every month to monitor changes in the bones from a long period without gravity. The ADUM team then worked with astronauts to perform additional ultrasound examinations on the station, including examinations of teeth, sinus and eyes. Finally a full examination of the heart was performed without direct video capabilities, using only voice guidance and relying on pattern recognition developed in the experiment.

More than 100 hours of ultrasound examinations were conducted with long-duration station crew members, providing a "head-to-toe" assessment of changes in the body associated with space flight. As of the end of 2009, the techniques and solutions developed by the team were officially accepted for medical support of the station crews as well as for conducting experiments in space physiology and clinical space medicine research.

Ultrasound Application Experience

ADUM investigators modified the training methods and remote guidance techniques developed for the space station to extend medical care capabilities on Earth. Non-physician athletic trainers for the Detroit Red Wings hockey team and the Detroit Tigers baseball team were taught advanced ultrasound skills to help with injured athletes. A portable ultrasound device was installed in the locker room of the athletic stadiums, and athlete-specific cue cards were developed for common sports injuries. Tele-ultrasound connections were established between the sporting arenas and Henry Ford Hospital in Detroit, Mich., to allow remote guidance capabilities.

Initial experiences with the hockey and baseball teams showed that the athletic trainers could perform complex muscle, bone and joint ultrasound examinations rapidly and with high diagnostic accuracy, allowing point-of-care diagnosis of athletic injury. Some ADUM investigators have extended these capabilities to the United States Olympic Training Facilities and have supported the Torino, Beijing and Vancouver games with hundreds of point-of-care ultrasound examinations in athletes with suspected injuries.

The ADUM team has extended the concept to remote environments, such as Mt. Everest and the Arctic Circle. The team designed a self-contained system that includes a portable ultrasound device, solar power, satellite phone connectivity and a laptop computer containing educational programs. An untrained mountaineer was able to perform a complete lung ultrasound scan at Advanced Base Camp on Mt. Everest using cue cards and remote guidance. The novice operator was able to send high-quality ultrasound images to the remote expert to diagnose a fellow climber with fluid in the lungs secondary to high altitude. A similar remote ultrasound system was used at Resolute Bay in the Canadian Arctic Circle to enable nonexpert operators to perform targeted scans of almost every organ system.

The just-in-time ultrasound educational programs developed for use in space and for remote locations also are appropriate for training health care personnel. A training program was developed for Wayne State University School of Medicine, and is now used by the American College of Surgeons to teach ultrasound to the surgeons of the future.

Future Plans

The ADUM team is currently developing a simple, whole-body ultrasound education catalog that can be used to teach nonexpert operators to diagnose a wide variety of conditions. An integrated remote medical care device is being created to combine computer education, ultrasound-based diagnostics and communication capabilities with remote experts to extend high-quality medical care capabilities to remote, rural and underserved regions on Earth.

The real-time nature of ultrasound imaging and the ability to easily transmit the images to allow remote expert guidance make ultrasound applications especially attractive for remote use. ADUM investigators are studying the possibility of using ultrasound to effectively answer primary clinical diagnostic questions in unconventional settings where ultrasound is the only (or the first available) source of imaging and where on-site expertise is limited. They have developed and extensively used multimedia e-learning software, which makes a winning combination with real-time mentoring of the distant on-site operator (remote expert guidance). These methods of focused ultrasound have been used successfully on the station and in several applications on the ground.

Ultrasound is one of the most adaptable diagnostic imaging modalities, which can be used for many medical and surgical conditions. Advances in portability and affordability, coupled with enhanced training programs and tele-ultrasound, can provide powerful diagnostic capabilities anywhere on and off the planet.

Together with its partners, the ADUM team will work to create new knowledge and capabilities to benefit human health on Earth and in the most daring exploration settings of the future.

Early Detection of Immune Changes Prevents Painful Shingles in Astronauts and Earth-Bound Patients[4]

Satish K. Mehta, Ph.D., Senior Scientist, Enterprise Advisory Services
Duane L. Pierson, Ph.D., Chief Microbiologist
C. Mark Ott, Ph.D., Senior Microbiologist
NASA Johnson Space Center

The physiological, emotional and psychological stress associated with spaceflight can result in decreased immunity that reactivates the virus that causes shingles, a disease punctuated by painful skin lesions. NASA has developed a technology that can detect immune changes early enough to begin treatment before painful lesions appear in astronauts and people here on Earth. This early detection and treatment will reduce the duration of the disease and the incidence of long-term consequences.

Spaceflight alters some elements of the human immune system: innate immunity, an early line of defense against infectious agents, and specific components of cellular immunity are decreased in astronauts. Astronauts do not experience increased incidence or severity of infectious disease during short-duration spaceflight, but NASA scientists are concerned about how the immune system will function over the long stays in space that may be required for exploration missions.

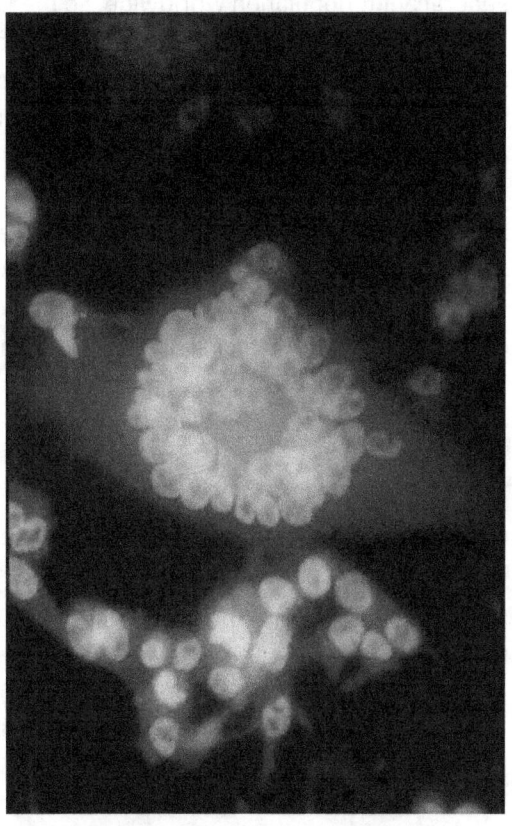

Selecting one or more biomarkers or indicators of immunity in healthy individuals is difficult, but the herpes viruses have become valuable tools in early detection of changes in the immune system, based largely on the astronaut studies. Eight herpes viruses may reside in the human body, and virtually all of us are infected by one or more of these viruses. Herpes viruses cause diseases including common "fever blisters" (herpes simplex virus or HSV), infectious mononucleosis (Epstein-Barr virus or EBV) and chicken pox and shingles (varicella zoster virus or VZV). In immune-suppressed individuals, herpes viruses may cause several types of cancer, such as carcinoma, lymphoproliferative disease and others.

VZV infected MeWo cells showing typical herpes-virus-induced multinucleated giant cells. Cultures are stained with acrydine orange to identify RNA (red) in the cytoplasm. (Image credit: NASA)

---

[4] Adapted from an original article that appeared in NASA Technology Innovation, Vol 15; 3, 2010; NP-2010-06-658-HQ.

According to the Centers for Disease Control and Prevention, one million cases of shingles occur yearly in the U.S., and 100,000 to 200,000 of these cases develop into a particularly painful and sometimes debilitating condition known as post-herpetic neuralgia, which can last for months or years. The other seven herpes viruses also exist in an inactive state in different body tissues much like VZV, and similarly they may also reactivate and cause disease during periods of decreased immunity.

The most common cause of decreasing immunity is age, but chronic stress also results in decreased immunity and increases risk of the secondary disease, such as VZV-driven shingles. Chemotherapy, organ transplants and infectious diseases, such as human immunodeficiency virus or HIV, also result in decreased immunity. Thus, viral reactivation has been identified as an important indicator of clinically relevant immune changes. Studies of immune-compromised individuals indicate that these patients shed EBV in saliva at rates 90-fold higher than found in healthy individuals.

The herpes viruses are already present in astronauts, as they are in at least 95 percent of the general adult population worldwide. So measuring the appearance of herpes viruses in astronaut body fluids serves as a much-needed immune biomarker. It is widely believed that various stressors associated with spaceflight are responsible for the observed decreased immunity. Researchers at NASA's Johnson Space Center found that four human herpes viruses reactivate and appear in body fluids in response to spaceflight. Due to the reduced cellular immunity, the viruses are allowed to emerge from their latent state into active infectious agents. The multiplying viruses are released into saliva, urine or blood and can be detected and quantified by a polymerase chain reaction or PCR assay for each specific virus. The PCR assay detects viral DNA and is very sensitive and highly specific, allowing the user to selectively replicate viral DNA sequences. The finding of VZV in saliva of astronauts was the first report of VZV being reactivated and shed in asymptomatic individuals. Subsequently, the VZV shed in astronaut saliva was found to be intact and infectious, posing a risk of disease in uninfected individuals.

PCR technology also was utilized to study VZV-induced shingles in patients from the general population. In the study, which merged the experience of physicians at the University of Texas Health Science Center in Houston and NASA scientists at Johnson Space Center, salivary VZV was present in each of the 54 shingles patients on the day treatment was initiated. Pain and skin lesions decreased following antiviral treatment, and levels of VZV decreased as well. The lower levels of VZV copies in shingles patients before treatment overlapped the upper range of VZV levels in astronauts, suggesting a potential risk of shingles from VZV reactivation in space.

This early-detection technology was used to facilitate early diagnosis of shingles in a 21-year-old patient, resulting in early medical intervention. This quick action prevented the formation of skin lesions and reduced the duration of pain associated with acute disease. In another NASA study, spaceflight PCR technology was utilized to detect VZV in the serum and peripheral blood mononuclear cells of all 25 shingles patients, demonstrating for the first time that viremia (virus in the blood) is a common manifestation of shingles. However, the PCR assay requires large, complex equipment, which is not practical for space flight.

To overcome this obstacle in an effort to investigate viral reactivation in crew members, NASA developed a rapid method of detection of VZV in body fluids, and a patent application is currently pending for it. The new technology requires a small sample of saliva, which is mixed with specialized reagents that produce a red color only when VZV is present. This technology makes possible early detection, before the appearance of skin lesions. Early detection allows for early administration of antiviral therapy and thus limits nerve damage and prevents overt disease. Rapid intervention is expected to prevent post-herpetic neuralgia. The device is designed for use in doctors' offices or spacecraft and can be modified easily for use with other viruses in saliva, urine, blood and spinal fluid. The sensitivity and specificity emanates from an antibody-antigen reaction.

In another collaborative study, NASA and University of Colorado Health Science Center (Denver) researchers developed a collection tool for assessment of salivary stress hormones during space shuttle missions. Saliva samples are collected on individual filter paper strips and dried. The dried samples remain stable at room temperature for as long as six months to allow testing once back on Earth. The test measures cortisol and dehydroepiandrosterone (DHEA), two important stress and immune regulatory hormones. The filter paper also can be used for proteins and other molecules of interest in saliva. Booklets of these filter papers now are being used in university and government laboratories for remote saliva collection.

These studies demonstrate the potential value of bringing to the general public a technology that could prevent a painful and debilitating condition in up to one million people each year in the U.S. alone.

Cancer Treatment Delivery

Tara Ruttley, Ph.D., Associate International Space Station Program Scientist
NASA

Humanity is on the constant search for improvements in cancer treatments, and the International Space Station has provided a microgravity platform that has enabled advancements in the cancer treatment process.

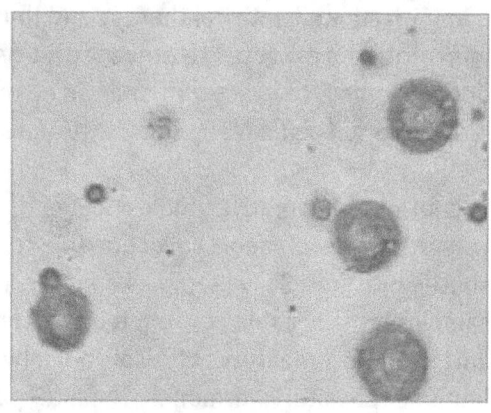

Single cell microencapsulation
(Image credit: NASA)

The oncology community has a recent history of using different microencapsulation techniques as an approach to cancer treatment. Microencapsulation is a single step process that forms tiny liquid-filled, biodegradable micro-balloons containing various drug solutions that can provide better drug delivery and new medical treatments for solid tumors and resistant infections. In other words, by using microcapsules containing antitumor treatments and visualization markers, the treatment can be directed right to the tumor, which has several benefits over systemic treatment such as chemotherapy. Testing in mouse models has shown that these unique microcapsules can be injected into human prostate tumors to actually inhibit tumor growth or can be injected following cryo-surgery (freezing) to improve the destruction of the tumors much better than freezing or local chemotherapy alone. The microcapsules also contain a contrast agent that enables C-T, X-ray or ultrasound imaging to monitor the distribution within the tissues to ensure that the entire tumor is treated when the microcapsules release their drug contents.

Dr. Morrison with MEPS flight hardware ready to pack for the International Space Station UF-2 mission

The Microencapsulation Electrostatic Processing System-II experiment, or MEPS-II, led by Dennis Morrison, Ph.D. (retired), at NASA Johnson Space Center, was performed on the station in 2002 and included innovative encapsulation of several different anti-cancer drugs, magnetic triggering particles, and encapsulation of genetically engineered DNA. The experiment system improved on existing microencapsulation technology by using microgravity to modify the fluid mechanics, interfacial behavior, and biological processing methods as compared to the way the microcapsules would be formed in gravity.

In effect, the MEPS-II system on the station combined two immiscible liquids in such a way that surface tension forces (rather than fluid shear) dominated at the interface of the fluids. The significant performance of the space-produced microcapsules as a cancer treatment delivery system motivated the development of the Pulse Flow Microencapsulation System, or PFMS, which is an Earth-based system that can replicate the quality of the microcapsules created in space.

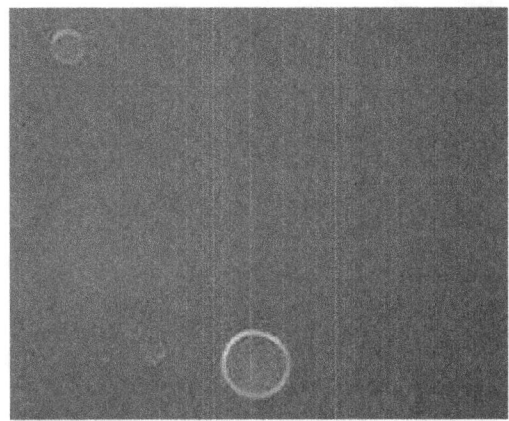

Microencapsulation containing anti-tumor drugs made on the space station. (Image credit: NASA)

As a result of this space station research, the results from the MEPS-II experiments have provided new insight into the best formulations and conditions required to produce microcapsules of different drugs, particularly special capsules containing diagnostic imaging materials and triggered release particles. Co-encapsulation of multiple drugs and Photodynamic Therapy, or PDT, drugs has enabled new engineering strategies for production of microcapsules on Earth designed for direct delivery into cancer tissues. Other microcapsules have now been made for treatment of deep tissue infections and clotting disorders and to provide delivery of genetically engineered materials for potential gene therapy strategies. Microcapsules that were made on the space station and are targeted at inhibiting the growth of human prostate tumors have been successfully demonstrated in laboratory settings. Although Morrison's team had performed several similar microencapsulation experiments on space shuttle missions, because of the space station's ability to support long-term experiments, more progress was made by the eight microencapsulation experiments conducted on the station in 2002 than from the 60+ prior experiments conducted on the four space shuttle missions—STS-77, STS-80, STS-95 and STS-107.

Benefits of Space Station Research

The microgravity environment on the station was an enabling environment that led the way to better methods of microcapsule development on Earth. The capability to perform sequential microencapsulation experiments on board the station has resulted in new, Earth-based technology for making these unique microballoons that provide sustained release of drugs over a 12–14 day period. The station research led directly to five U.S. patents that have been licensed by NASA and two more that are pending. NuVue Therapeutics, Inc., is one of several commercial companies that have licensed some of the MEPS technologies and methods to develop new applications, such as innovative ultrasound enhanced needles and catheters that will be used to deliver the microcapsules of anti-tumor drugs directly to tumor sites. More recent research uses a new device for freezing tumors ("cryo-ablation") followed by ultrasound-guided deposition of the multi-layered microcapsules containing different chemotherapy drugs outside the freeze zone within a human prostate or lung tumor. In a 28-day study, combination therapy resulted in retarding tumor growth 78 percent and complete tumor regression of up to 30 percent after only three weekly injections of microencapsulated drug at tiny quantities that should not have slowed down tumor growth by more than 5–10 percent. NuVue Technologies, Inc., has now obtained two U.S. patents based on the combination therapy that includes the delivery of the NASA-type microcapsules. Upon securing funding, clinical trials to inject microcapsules of anti-tumor drugs directly into tumor sites will begin at MD Anderson Cancer Center in Houston and the Mayo Cancer Center in Scottsdale, Ariz.

Other potential uses of this microencapsulation technology include microencapsulation of genetically engineered living cells for injection or transplantation into damaged tissues, enhancement of human tissue repair, and real-time microparticle analysis in flowing sample streams that would allow petrochemical companies to monitor pipeline volume flow.

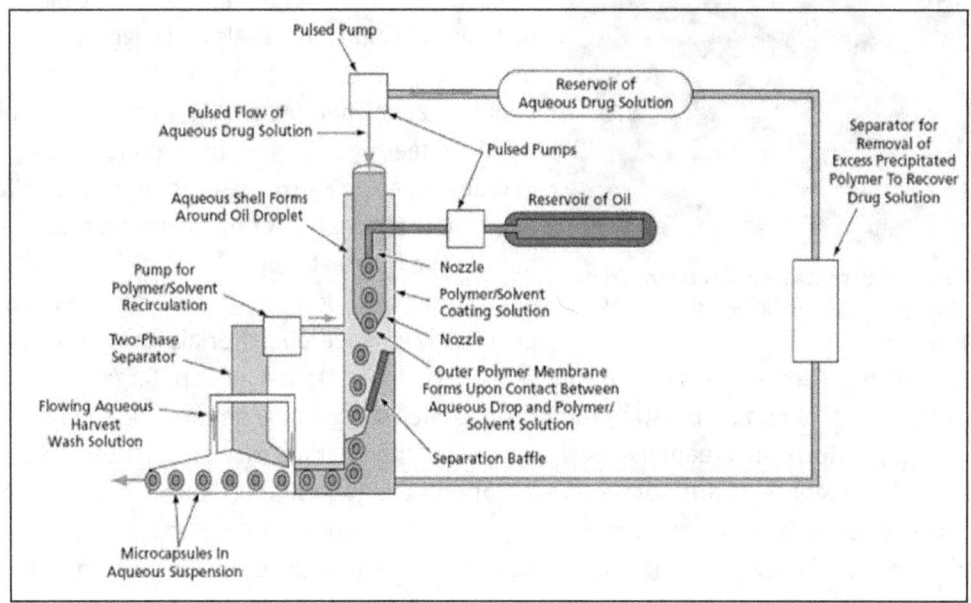

Schematic of the Pulse Flow Microencapsulation System (Image credit: NASA)

Advanced NASA Technology Supports Water Purification Efforts Worldwide

Arun Joshi, International Space Station Program Science Office
NASA Johnson Space Center

Whether in the confines of the International Space Station or a tiny hut village in sub-Saharan Africa, drinkable water is vital for human survival. Unfortunately, many people around the world lack access to clean water. Using technology developed for the space station, at-risk areas can now gain access to advanced water filtration and purification systems, making a life-saving difference in these communities.

The nonprofit organization Concern for Kids, or CFK, in Nye, Mont., has provided aid and disaster relief to countries such as Malaysia, Iraq and Indonesia since 1992. Among other services, the group raises funds to install water delivery systems and water storage tanks in at-risk regions.

Years later, CFK representatives learned about a deep-water well failure in the tiny Kurdish village of Kendala, Iraq, which left its residents without access to drinkable water. The population quickly dwindled from 1,000 residents to a mere 150. Those remaining were forced to use a nearby creek that contained water contaminated by livestock, which they sifted through fabric to remove dirt and debris.

Todd Harrison was president of CFK's board of directors at the time and strongly empathized with the people of Kendala. He set out on a mission to revive this ailing community by improving the deplorable conditions. The solution came in the form of a familial connection that put Harrison in touch with NASA engineers who developed technology to provide clean water aboard the space station.

Harrison's sister, Robyn Carrasquillo, was the engineering manager for the Environmental Control and Life Support System, or ECLSS, project at NASA's Marshall Space Flight Center. She and her team of engineers were responsible for developing the cutting-edge water purification system that recycles air and water aboard the station.

By efficiently recycling wastewater aboard the space station, there is a reduced need to provide the resource via resupply—which would not be an option for long-duration space travel. Without this capability, the station's current logistics resupply capacity would not be able to support the standard population of six crew members. "In a year, the water saved by recycling exceeded the initial launch weight and weight of replacement parts due to some initial issues," commented Carrasquillo.

Two principal components make up the International Space Station Regenerative ECLSS: The Water Recovery System, or WRS, and the Oxygen Generation System, or OGS. The WRS conducts the water purification and filtration process in the ECLSS. Commercial companies took an interest in this part of the ECLSS project, as they sought to adapt it to an Earth-based water treatment system.

The International Space Station water recovery component of the Environmental Control and Life Support System (NASA image)

Harrison discovered an interesting relationship between CFK's water filtration system and NASA because of his familiarity with his sister's work. NASA's previous research and application provided the Microbial Check Valve, or MCV, an integral component of the purification and filtration process.

Volunteers help install and test a water purification system in Kendala, Iraq. (Image courtesy of Concern for Kids)

Originally developed for the Space Shuttle Program by Umpqua Research Corporation, the MCV was later sold to WSC. The MCV is an iodinated-resin that provides a simple way to control microbial growth in water without the use of power. By dispensing iodine into the water, it performs an important secondary nutritional function for the populace. This chemical, when added to the diet, promotes proper brain functions and maintains bodily hormone levels—which regulate cell development and growth. Children who lack iodine in their diets exhibit growth mental retardation.

With the help of U.S. Army Civil Affairs and Psychological Operations Command (Airborne) personnel, a 2,000-liter water tank and fresh water were brought to the Kurdish village. Workers ensured that the water was clean and iodinated to prevent bacteria and virus contamination. Carrasquillo recalls the installation efforts and associated challenges. "[CFK] encountered some technical issues that our group helped with by phone. The MCV had dried out during transit and we were able to do an assessment and reassure them it should be okay. In addition, the pumps that were available in the local village were oversized for the filtration system."

Carrasquillo's team strategized and implemented a bypass that allowed workers to use the available pump and start the system immediately. This workaround enabled the successful processing of Kendala's water supply.

Joint collaborations between aid organizations and NASA technology show just how effectively space research can adapt to contribute answers to global needs. Since this initial effort, the commercialization of this station-related technology has provided aid and disaster relief for communities worldwide. Chiapas, Mexico; Kampang Salak, Malaysia; Sabana San Juan, Dominican Republic; Balakot, Pakistan; and Vera Cruz, Mexico are just a few examples of cities that benefited from this advanced filtration capability. Other related developments include a personalized device that uses forward osmosis to turn urine into drinking water. Water Security Corporation, or WSC, licensed the device for use in commercial ground-based filtration systems deployed around the world.

Earth Observation and Disaster Response

Earth Remote Sensing from the Space Station—It's Not Just Handheld Cameras Anymore

William L. Stefanov
Chief Scientist, Science Applications, Research and Development Department, Jacobs[5]

Since the International Space Station became operational in November 2000, astronauts on board have taken more than 600,000 images of the Earth's land surface, oceans, atmospheric phenomena, and even images of the Moon from orbit using handheld film and digital cameras as part of the Crew Earth Observations experiment. Despite this large volume of imagery and clear capability for Earth remote sensing, the space station historically has not been perceived as an Earth observation platform by many remote sensing scientists. With the installation of new facilities and sophisticated sensor systems on the International Space Station over the past two years—and more to come—this perception is changing.

So what can the station offer in terms of Earth remote sensing that free-flying, robotic satellite systems cannot?

Images With a Variety of Lighting Conditions

Unlike many of the traditional Earth observation platforms, the station orbits the Earth in an inclined equatorial orbit that is not sun-synchronous. This means that the station passes over locations on the Earth between 52 degrees north and 52 degrees south latitude at different times of day and under varying illumination conditions. Remote satellite-based Earth observing sensors are typically placed on polar-orbiting, sun-synchronous platforms like Landsat7 or Terra in orbits designed to pass over the same spot on the Earth's surface at approximately the same time of day. These satellite platforms will revisit a location

ISSAC "first light" image of Charlotte Harbor, Fla. acquired June 10, 2011, overlaid on Landsat 5 base image. The ISSAC scene is processed to highlight vegetation in red, urban areas in gray, and water surfaces in black. Clouds appear bright white.

[5] William L. Stefanov is with Jacobs, the prime contractor for the NASA Johnson Space Center Engineering and Science Contract.

about every two weeks. While collecting imagery in similar lighting conditions is good for producing uniform data for a specific place, it restricts the time that data is collected (near local solar noon in most cases). If a scientist is interested in a surface process that typically happens in the early morning or late in the afternoon (for example, patterns in coastal fog banks), the data will be difficult to collect from the polar-orbiting, sun-synchronous satellites.

Responsive Data Collection

Another advantage unique to the space station is the presence of crew that can react to unfolding events in real time, rather than needing a new data collection program uploaded from ground control. This is particularly important for collecting imagery of unexpected natural hazard events such as volcanic eruptions, earthquakes, and tsunami. The crew can also determine whether viewing conditions—like cloud cover or illumination—will allow useful data to be collected, as opposed to a robotic sensor that collects data automatically without regard to quality.

Complementing handheld, digital camera imagery taken by astronauts, the current automated sensor systems and facilities on board the space station—both internal and external—provide exciting new capabilities for Earth remote sensing. In addition, the station's power and data infra-structure encourages the development of new sensors. The following systems managed by NASA for Earth observation are now or will shortly be onboard and operational for the space station (or manifested for transport):

- Window Observational Research Facility, or WORF, provides a highly stable internal mounting platform to hold cameras and sensors steady while offering power, command, data, and cooling connections. With the WORF, the high-quality optics of the nadir viewing window—looking "straight down" at the Earth—in the U.S. Destiny Laboratory are now fully utilized for the first time.

- International Space Station Agricultural Camera, or ISSAC. The ISSAC was developed by students and faculty at the University of North Dakota. Its prime purpose is to collect

HICO true-color image of Monterey Bay, Calif., acquired March 27, 2010.

multispectral data supporting agricultural activities and related research in the Upper Midwest of the United States. In addition, the ISSAC can be tasked to collect imagery of natural hazards and disasters in support of NASA humanitarian efforts. ISSAC collects information in the visible and near-infrared wavelengths (3 bands) at a nominal ground resolution of 20 meters per pixel.

- Hyperspectral Imager for the Coastal Ocean, or HICO, is mounted on the external facility of the Japanese Kibo module. The prime mission of HICO is to collect data on water clarity, bottom materials, bathymetry, and on-shore vegetation along the coasts of Earth's oceans at approximately 90 meters per pixel ground resolution. The sensor collects high-quality information in 87 bands over the visible and near-infrared wavelengths.

- International Space Station SERVIR Environmental Research and Visualization System, or ISERV, is a planned sensor system consisting of a Schmidt-Cassegrain telescope paired with a digital camera system to collect visible-wavelength imagery at ground resolutions of less than 3 meters per pixel. The system will be mounted in the WORF and will have excellent stability and targeting capabilities. The sensor is intended to support the SERVIR program in its goals of humanitarian support by using Earth science data to aid the developing world, mitigate and respond to disasters, and provide humanitarian support.

Currently, individual science teams manage access to data collected by the various sensor systems, but a central data access facility is planned for the future. The combined capabilities of both human-operated and autonomous sensor systems onboard the space station promise to significantly improve our ability to monitor the Earth and respond to natural hazards and catastrophes. Integration of the space station Earth observation systems represents a significant and complementary addition to the international satellite-based Earth observing "system of systems," providing knowledge and insight into our shared global environment.

Space Station Imagery Helps Island Nations Manage Coral Reef Resources

NASA Johnson Space Center

Coral reefs are a critical resource for many island and coastal nations, particularly in the tropical Pacific and Indian Oceans. These nations rely on the rich reef ecosystem to support local fisheries; additionally, the reefs serve as barriers that help protect coasts from significant waves and storm surges and bring in substantial revenue through tourism. Astronaut photographs of coral reefs taken from the International Space Station provide an important perspective on coral reef geography, coastal development, and related land habitats.

In recent years, the scientific community has recognized that reefs around the world are threatened by large scale processes such as global warming and associated sea level rise, ocean warming and acidification, and local human activities such as coastal development, over-fishing and heavy tourism (see Reefs at Risk and Mapping the Decline of Coral Reefs). Collapse of individual reef systems would spell disaster for local island nations and coastal societies dependent on reef resources. A global decline in reef systems carries significant implications for the oceans' biodiversity, support for important fisheries, and more.

About a decade ago, several programs were initiated to map the extent of global reefs as a first step in understanding their geographic attributes and managing the resource (see Millennium Coral Reef Mapping). As these initiatives began, scientists recognized that 1) accurate maps of the world's reefs were lacking; 2) detailed mapping of the reefs would provide more value to local resource managers; and 3) the remoteness of many reef systems would make detailed mapping difficult or impossible.

Most reef data came from established remote sensing sources, including Landsat7 and sea-viewing Wide Field-of-View Sensor, or WiFS. However, teams of scientists and resource managers quickly recognized that very little detailed imagery exists over many remote oceanic locations. NASA supported the collection of reef imagery from a variety of sensors through partnership with several organizations (e.g., United Nations Environmental Program–World Conservation Monitoring Centre and International Coral Reef Action Network, International Center for Living Aquatic Resources Management, and ReefBase).

Astronaut images were identified as an important contributing resource because they provided better resolution than some remotely-sensed data, were generally cloud-free, and covered areas not traditionally imaged by remote sensing platforms. Astronauts consciously target beautiful reef areas as they orbited over the large expanses of oceans. Although not a prime source of remotely sensed data, astronaut photography of remote reef locations in the world's oceans have provided valuable data that have been assimilated and fused into global databases that help to map reef systems, and monitor their changes. This information is critical for local reef managers.

Today, station astronauts routinely collect high resolution (5–6 m) digital imagery of the Earth's reefs. Richard Stumpf of the National Oceanic and Atmospheric Administration or NOAA reported in a 2003 study that the images have been used by NOAA for water depth assessments. Additional collaborations have evaluated the influence of the spatial scale of the imagery on determining landscape parameters of atolls as reported by Serge Andréfouët of the Institute for Marine Remote Sensing, University of South Florida, in 2003, and scientists use the high resolution astronaut photography for planning dive surveys and other management activities according to a 2002 publication by Jean-Pascal Quod of the French initiative for Coral Reefs, or IFRECOR.

Photographing reef areas from the station is an established way to contribute to the global database of reef images and support coastal communities in managing their own reef areas. In the future, other space station sensors, such as the Hyperspectral Imager for the Coastal Ocean, or HREP-HICO, and International Space Station SERVIR Environmental Research and Visualization System, or ISERV, may also provide meaningful data for reef mapping activities.

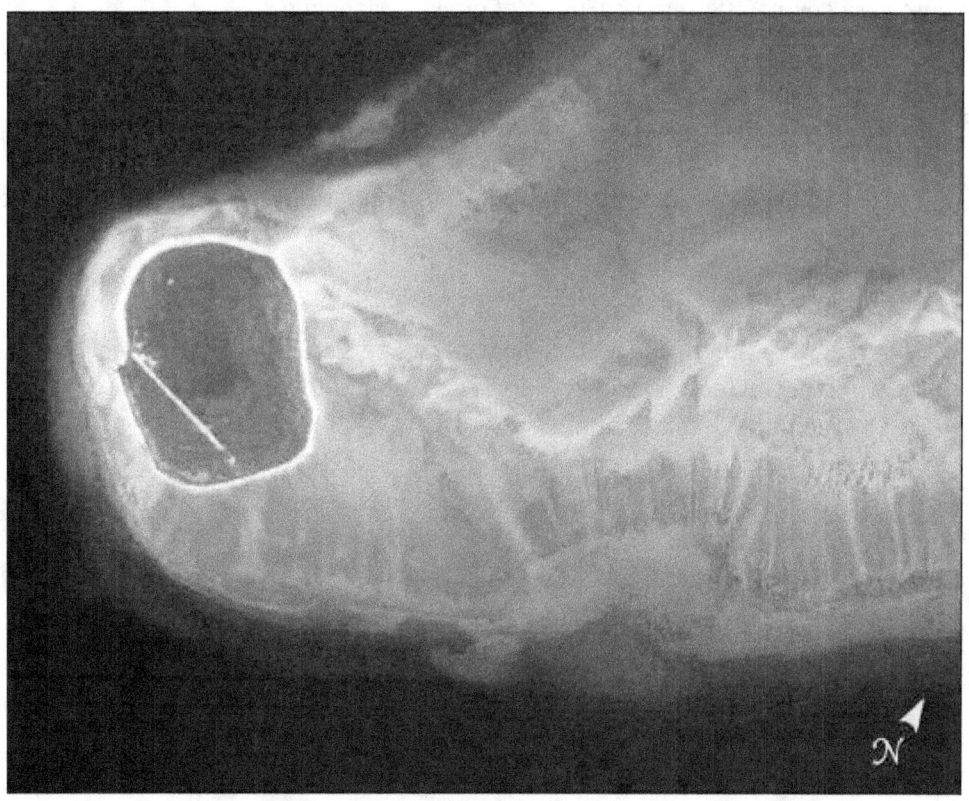

Îles Glorieuses (îles Eparses archipelago, Indian Ocean). These islands are protected because of their importance for sea turtles and seabird nesting. (ISS002-E-6913)

The digital photograph above, taken from the space station, shows details of the reefs surrounding îles Glorieuses. The photograph served as a base map for detailed field mapping of the reef zones (geomorphology and ecology) by Agence pour la Recherche et la Valorisation Marines (ARVAM), based in Réunion, and partners. These maps supported biological inventories and surveys of coral reef health using established international protocols (Global Coral Reef Monitoring Network and Reef Check). The high spatial resolution of these astronaut photographs (about 5 meters per pixel) captures the detail needed to support this kind of detailed field research.

References

Andréfouët, S., J. A. Robinson, C. Hu, G. Feldman, B. Salvat, C. Payri, and F.E. Muller-Karger. 2003. Influence of the spatial resolution of SeaWiFS, Landsat 7, SPOT and International Space Station data on determination of landscape parameters of Pacific Ocean atolls. Canadian Journal of Remote Sensing 29(2):210–218.

Robinson, J. A., G. C. Feldman, N. Kuring, B. Franz, E. Green, M. Noordeloos, and R. P. Stumpf. 2000. Data fusion in coral reef mapping: working at multiple scales with SeaWiFS and astronaut photography. Proceedings of the 6th International Conference on Remote Sensing for Marine and Coastal Environments, Vol. 2, pp. 473–483.
http://eol.jsc.nasa.gov/newsletter/DataFusionInCoralReefMapping/

Spalding, M. D., C. Ravilious, and E. P. Green. World Atlas of Coral Reefs. University of California Press, Berkeley.

Stumpf, R. P., K. Holderied, J. A. Robinson, G. Feldman, N. Kuring. 2003a. Mapping water depths in clear water from space. Coastal Zone 03, Baltimore, Maryland, 13–17 July 2003
http://eol.jsc.nasa.gov/newsletter/CoastalZone/default.htm

Bryant, D., L. Burke, J. McManus, and M. Spalding. 1998. Reefs at Risk: A map-based indicator of threats to the world's coral reefs. World Resources Institute, Washington, D.C.
http://wri.igc.org/reefsatrisk/reefrisk.html

Andréfouët, S., J. A. Robinson. 2003. The use of Space Shuttle images to improve cloud detection in tropical reef environment. International Journal of Remote Sensing 24: 143–149.

Space Station Agricultural Camera Observes Flooding in North Dakota

William L. Stefanov
Science Applications, Research and Development Department, Jacobs

Ask a geologist or ecologist about remote sensing of the Earth—collecting information about a material or process without physically touching or sampling it—and they will likely mention Landsat; Advanced Spaceborne Thermal Emission and Reflection Radiometer, known as ASTER; Moderate Resolution Imaging Spectroradiometer, or MODIS; or other sensor systems that routinely collect data from orbit for the land surface, oceans, or atmosphere. Since the early 1970s, terrestrial global remote sensing has been the territory of the polar-orbiting, sun-synchronous satellite, but that is now changing with the addition of new remote sensing systems to the International Space Station.

One of the first new systems is the International Space Station Agricultural Camera, or ISSAC. The ISSAC was developed by students and faculty at the University of North Dakota, and its prime purpose is to collect data in support of agricultural activities and related research in the upper midwest of the United States. Today's farmers use data and information from a variety of remote sensing satellites to understand weather systems and climate patterns and to monitor the health of their crops. The ISSAC will be another source of information for farmers and agricultural researchers by collecting imagery of croplands and other land cover over the Midwestern states during the growing season.

Unlike the more familiar handheld digital camera photographs taken by astronauts from the space station for the Crew Earth Observations experiment, ISSAC imagery is similar to multispectral data collected by the ASTER sensor onboard the NASA Terra satellite. The ISSAC is also mounted in the Window Observational Research Facility, or WORF, and operates independently of the station crew.

The sensor system uses two digital still cameras equipped with filters to collect individual image frames sensitive to the visible green, visible red, and near-infrared wavelengths of the electromagnetic spectrum. The individual frames are then combined to form a single multispectral (three-band) image. This wavelength combination makes the ISSAC particularly good at distinguishing different kinds of vegetation and detecting changes in the areal coverage and health of plants, two factors of importance for agricultural studies and monitoring.

In addition, the ISSAC can collect imagery of natural hazards and disasters in support of NASA humanitarian efforts. ISSAC demonstrated this capability soon after it became operational by collecting images of flooding from the Souris River near Minot, N.D., on June 24, 2011. Because ISSAC imagery is recorded in wavelengths similar to Landsat, scientists could easily compare the ISSAC data with earlier Landsat data of the Souris River valley, clearly illustrating the extent of flooding in and around Minot.

With the ISSAC now operational, data collected from the space station includes imagery that is similar to that acquired by polar-orbiting, satellite-based sensors. ISSAC significantly augments the traditional Earth observations from polar-orbiting satellites by collecting data at variable times of day and with different repeat frequencies. Addition of the ISSAC and other remote sensing systems to the station, whether housed internally in the WORF or mounted externally, improves the capability of NASA to observe and monitor natural and anthropogenic (of human origin) Earth processes and natural hazards.

Comparison of Minot, N.D., and Souris River Valley during normal river flow conditions (top image, Landsat Thematic Mapper data) and during flood conditions (bottom image, ISSAC data). Both images have been processed to highlight actively photosynthesizing vegetation in red. Urban areas appear gray-brown and water-flooded.

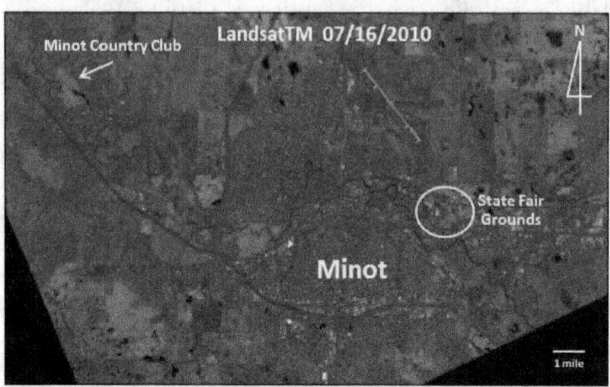

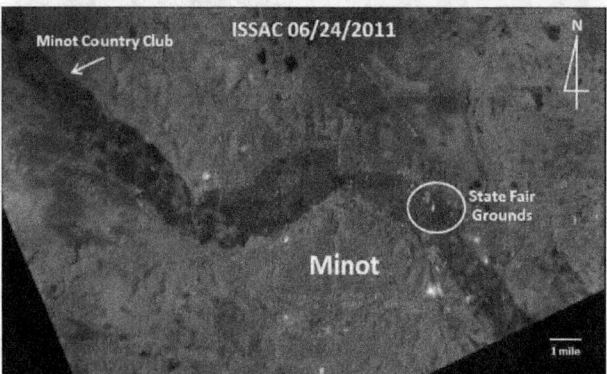

Monitoring the Health of the Lagoon of Venice From the Space Station

William L. Stefanov
Science Applications, Research and Development Department, Jacobs

The City of Venice, Italy, is known for its architecture, history, romance, and, of course, the canals that serve as major thoroughfares through the urban area. The canals and bridges are the human overprint on the Lagoon of Venice, a marshy body of water along the Adriatic Sea that contains the 117 islands upon which the city is built. In addition to forming the base of the city of Venice, the Lagoon is a critical part of the Mediterranean wetlands ecosystem, recognized by international treaty as a Ramsar site.[6]

Like many coastal regions, the Lagoon of Venice, together with the city of Venice and the environmentally sensitive wetlands, is subsiding into the Adriatic Sea, creating flood risks for Venice and major damage to the wetlands. Careful mapping of the Lagoon is critical to mitigating damage to the people and environment.

In 2010, astronaut photographs of the Lagoon taken from the International Space Station caught the eye of Dr. Alessandro Mulazzani, an environmental consultant working with the Atlante della laguna (Atlas of the Lagoon) project (http://www.silvenezia.it/). The Atlas is a web-based Geographic Information System (GIS) that provides a wealth of information on the climate, ecosystem, hydrology, and human impacts to the Lagoon. Mulazzani characterized the value of high-resolution astronaut photography to the Atlas: "These photographs are very useful for us to provide an up-to-date view of this unique environment, the Lagoon of Venice, especially for general public and students…"

This interest developed into an ongoing collaboration with the Crew Earth Observations, or CEO, team at NASA's Johnson Space Center and resulted in the Lagoon of Venice becoming an official CEO site for continued astronaut photography. The CEO team uses the most up-to-date station orbital information to predict when particular ground sites will be visible to the crew. These target predictions are then screened for adequate lighting, acceptable degrees of cloud cover, and the availability of the crew to take imagery. Ground sites that pass through these various filters are provided to the station crew members as targets for imaging using the cameras aboard the station. Following successful acquisition of imagery for a requested target, the images are cataloged and added to the Gateway to Astronaut Photography of Earth online database and become freely available to the international public. For formal collaborations, such as the Lagoon of Venice, notification of new imagery can also be sent to designed investigators.

The use of handheld digital camera imagery from the station can help fill the gaps in time series of robotic satellite data as well as provide imagery taken from different viewing perspectives and

---

[6] The Ramsar Convention, first organized in 1971 in Ramsar, Iran, is an international treaty intended to foster maintenance and sustainable use of the wetlands within the member nation's territories. The Lagoon of Venice became a Ramsar site in 1989.

ground resolutions. This broad range of performance is a unique advantage of the station as a remote sensing platform and particularly highlights the advantage of having a human-operated remote sensing capability in addition to the more traditional autonomous systems. In the case of the Lagoon of Venice, having human eyes watching for unexpected opportunities to collect data may help ensure the health of this wetland for generations to come.

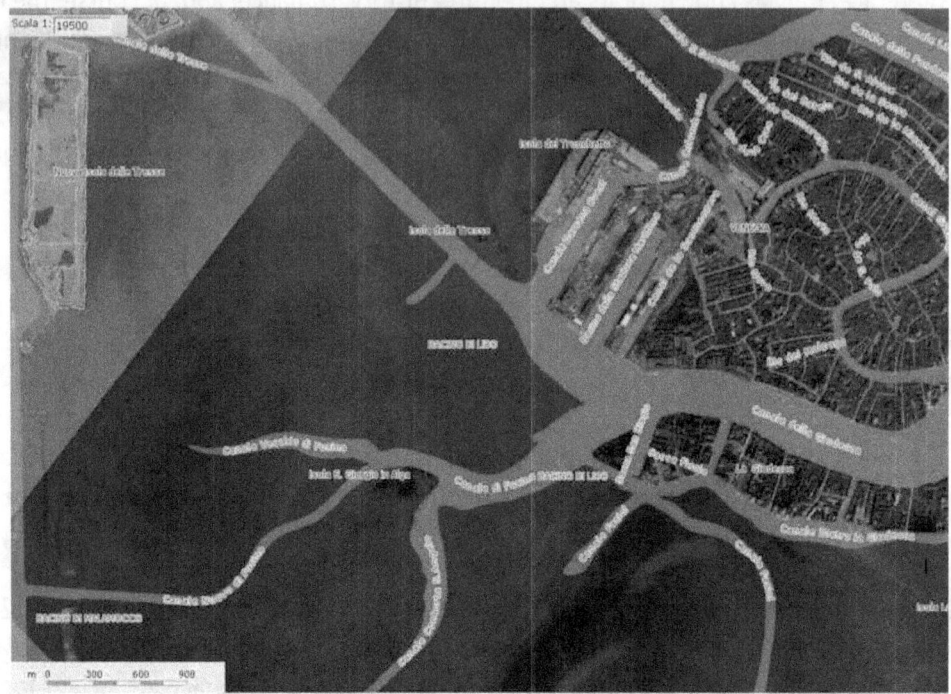

Screen capture of the Atlas of the Lagoon, showing geo-referenced astronaut photograph ISS023-E-13766 with a portion of the canal system (bright blue) added as an additional layer. The base image is an orthophoto mosaic (image created from a series of geometrically corrected images). Note that the site is currently only available in Italian.
(http://www.silvenezia.it/webgis/map.phtml?config=baseorto)

Unparalleled Views of Earth's Coast With HREP-HICO

Arun Joshi, International Space Station Program Science Office
NASA Johnson Space Center

Scanning the globe from the vantage point of the International Space Station is about more than the fantastic view. While cruising in low Earth orbit, the space station HICO and RAIDS Experiment Payload–Hyperspectral Imager for the Coastal Ocean, or HREP-HICO, gives researchers a valuable new way to view the coastal zone.

Using an imaging spectrometer mounted outside the station on the Japanese Exposed Facility of the Kibo Laboratory, researchers are collecting data about the Earth that will help them to better understand coastal environments and other regions around the world.

Why is this important? Coastal waters are an important link between local and global economic development and environmental sustainability. Coastal zones support many of the world's major cities (and their industrial zones, ports, recreational facilities); they also include critical ecosystems that support fisheries and protect shorelines.

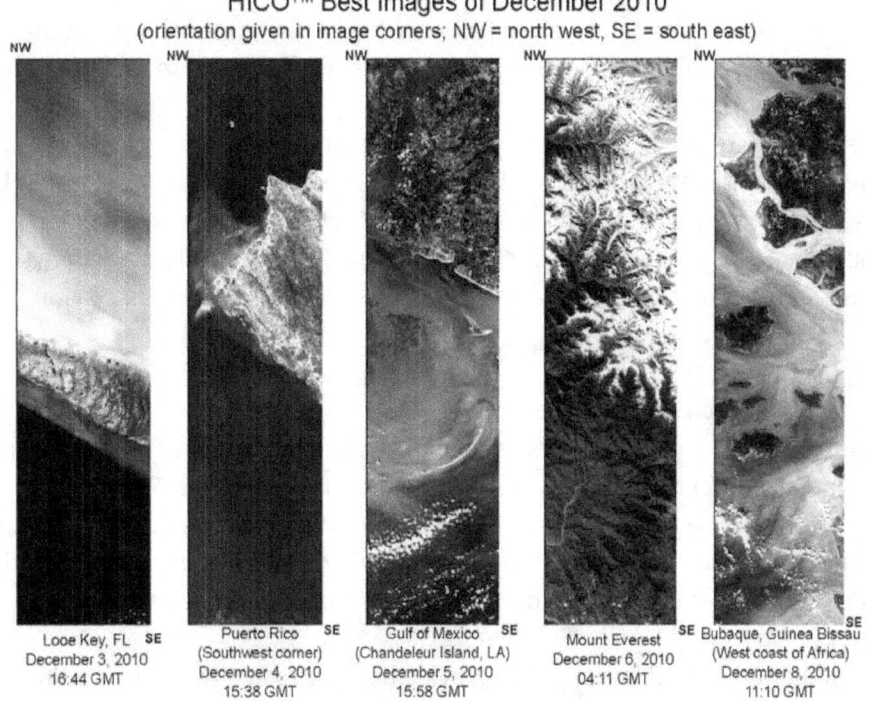

HICO™ Best Images of December 2010
(orientation given in image corners; NW = north west, SE = south east)

Looe Key, FL SE
December 3, 2010
16:44 GMT

Puerto Rico SE
(Southwest corner)
December 4, 2010
15:38 GMT

Gulf of Mexico SE
(Chandeleur Island, LA)
December 5, 2010
15:58 GMT

Mount Everest SE
December 6, 2010
04:11 GMT

Bubaque, Guinea Bissau SE
(West coast of Africa)
December 8, 2010
11:10 GMT

The above image compilation is an annotated representation of the best pictures taken during the December 2010 investigation of HICO. (Image courtesy of NASA)

HREP-HICO is a visible to near-infrared wavelength spectrometer optimized for environmental characterization of coastal zones and mapping of terrestrial geophysical features of the Earth. Mike Corson, Ph.D., principal investigator for HREP-HICO at the Naval Research Laboratory in Washington, D.C., explained that "Coastal environmental characterization includes producing maps of near-shore bathymetry, water clarity, organic and inorganic dissolved and suspended matter and bottom characteristics…. Such maps are important to Navy and Marine Corps who operate in coastal areas."

The Payload Operations and Integration Center at the Marshall Space Flight Center in Huntsville, Ala., manages operations for HREP-HICO. Aside from the initial installation and eventual removal from the station via extravehicular robotics, the collection of data from HREP-HICO requires no crew interaction.

Taking images of the ocean, however, is harder than it seems. As the first space-borne imaging spectrometer designed to sample the coastal ocean, HREP-HICO has to account for the vast breadth of global coastlines, the complex makeup of the ocean, changing weather conditions, and the location of the sun. These challenges can easily obscure views of the water and the sea floor—yielding poor and incomprehensible imagery.

HREP-HICO records reflected light from a range of wavelengths—including visible and near-infrared wavelengths. This spectral information is used to identify and quantify each pixel in the image scene. This allows for subsequent correction for atmospheric effects and sea surface reflections.

HREP-HICO also demonstrates its utility to meet Department of Defense requirements by validating the performance of Maritime Hyperspectral Imaging, or MHSI, technology. MHSI technology provides imagery performance requirements and product retrieval algorithms tailored by scientists for the coastal zone. "Product algorithms for MHSI often make use of biological or physical information relevant to the imaged area because of the complicated nature of coastal environments," Corson explained.

The Office of Naval Research sponsors the HREP-HICO instrument as an Innovative Naval Prototype. The HICO prototype has two goals. The first is to demonstrate the ability to produce maps of coastal environmental properties using hyperspectral imagery from space. The second is to demonstrate ways to reduce the cost and schedule of building a space payload. By manipulating design and using commercial-off-the-shelf components where possible, engineers effectively designed HREP-HICO at a fraction of the cost of traditional space instruments.

The results from the HREP-HICO investigation provide benefits to agencies with marine responsibilities, such as the National Oceanic and Atmospheric Administration, or NOAA, via information on bathymetry, bottom type, water clarity, and other water optical properties. This technology can also aid in land studies for agricultural purposes by monitoring land cover, vegetation type, vegetation stress and health, and crop yield. Researchers may also use HREP-HICO to determine the environmental impacts of natural and unnatural disasters, such as the 2010 oil spill in the Gulf of Mexico.

The HREP-HICO launched Sept. 10, 2009, and was mounted to the International Space Station's Japanese Experiment Module–Exposed Facility, or "front porch." HREP-HICO is the first U.S. experiment payload mounted on the Exposed Facility.

HREP-HICO collected its first imagery on Sept. 25, 2009. According to Corson, "HICO has recorded almost 4,000 hyperspectral images of coastal environments worldwide during its ongoing mission. HICO will continue its pathfinder missions, providing image data to dozens of government and university researchers."

Space Station Captures Tsunami Flooding in Northern Japan

William L. Stefanov
Science Applications, Research and Development Department, Jacobs

On Mar. 11, 2011, the eastern coast of Japan was shaken by the magnitude 9.0 Tohoku earthquake—one of the strongest earthquakes ever recorded. Caused by fault movement between the Pacific and North American tectonic plates, the quake spawned a tsunami that inundated much of the eastern coastline of the island of Honshu. Along with the tragic loss of human life and widespread devastation to buildings, infrastructure, and agriculture, the tsunami damaged the Fukushima Daiichi nuclear power station, leading to radiation leaks and potentially long-term ecological hazards.

The crew of the space station responded to the crisis and acquired several useful images of flooding from the tsunami. This image, taken on Mar. 13, 2011, illustrates flood waters along the Japanese coast north and east of Sendai. An astronaut on the International Space Station took this image of Higashimatsushima from an altitude of 220 miles (350 kilometers). Both agricultural fields and settled areas are submerged by muddy water, while the crisscrossing runways at Matsushima Airport are surrounded.

Oblique image of the Japanese coastline north and east of Sendai following inundation by a tsunami. The photo was taken Mar. 13, 2011. Sunglint indicates the widespread presence of floodwaters and indicates oils and other materials on the water surface.
(http://eol.jsc.nasa.gov/EarthObservatory/Tsunami_Japan_2011_glint.htm)

Far from being dispassionate observers of the disaster, the station crew members expressed their concern and sympathy for the people of Japan via video, radio, and email exchanges. At the onboard press conference on Apr. 13, Cosmonaut Alexander Samokutyaev said, "The tragedy that shocked everybody across the world—the horrible natural disaster that struck the island nation of Japan caused a lot of pain in every Russian's heart. Since we are neighboring nations and really close to each other and this tragedy can easily happen and impact our territory as well. On our behalf, I would like to say that we would like to enhance the monitoring capabilities. Perhaps we'll have some new capabilities using scientists all over the world to track and monitor natural disasters in a new fashion that—God forbid—should happen again anywhere in the world."

Astronaut Ron Garan said, "We wanted to say that the strength of our international partnership lies in the fact that we are all together, in the good times and the bad times, and we support each other." The Mar. 11 earthquake caused severe damage to oil refineries, some of which caught fire. In the aftermath, oil floated on the surface of Ishinomaki Bay. In the photo below, sunglint—the mirror-like reflection of the sun on the ocean's surface—highlights the oil slicks; oil smoothes the surface and makes the water more reflective. In this image, patches of oil tend to appear lighter than oil-free areas. Other phenomena, however, can lighten the water's appearance, especially close to the shore.

The image illustrates two unique aspects of the station for Earth observation and disaster response. Using handheld digital cameras, the crew has the capability to capture sunglint on water surfaces with greater frequency and control than most satellite-based systems—this provides an enhanced capability to detect and map standing water on the land's surface and may also indicate areas of particular environmental and health concern for contaminants. True-color images such as these, taken at a variety of pixel resolutions, can also be transmitted to responders on the ground and are readily understandable—little to no post-processing of the imagery is required.

As an Earth remote-sensing platform, the station is unique in that it hosts both human-operated and autonomous sensor systems. The "human element" has proved to be useful in both recognizing and capturing imagery of natural processes and features—including catastrophic events—and in providing views that complement other imagers, extending the usefulness of the remotely sensed data.

Superconducting Submillimeter-Wave Limb-Emission Sounder (SMILES)

Takuki Sano, International Space Station Science Project Office
Institute of Space and Astronautical Science, JAXA

The Superconducting Submillimeter-Wave Limb-Emission Sounder (SMILES) is the first onboard mechanically cooled superconducting mixer and high-resolution system for measuring atmospheric minor constituents related to stratospheric and mesospheric chemistry. SMILES collected high-quality observation data for six months until the instrument encountered trouble. These data indicate the excellent performance of the SMILES instrument as the data analysis progressed, and the data are expected to produce a mine of new scientific information.

The main scientific objective of the SMILES mission is to study the recovery and stability of the stratospheric ozone, also known as the ozone layer. Some numeric atmospheric model calculations have suggested that global ozone levels will recover to pre-1980 levels around the middle of the 21st century. However, there are still considerable uncertainties affecting ozone levels, especially in the bromine budget and chemical processes of inorganic chlorine. In addition, stratospheric cooling due to increases in greenhouse gases in the troposphere may affect the trend and the stability of the ozone layer. The SMILES mission contributes to these studies by focusing on the detailed halogen chemistry related to ozone destruction processes.

SMILES was launched to the International Space Station on Sept. 11, 2009, with Konotori1 on the HIIB launch vehicle from Tanegashima Space Center in Japan. After being attached to a port of the Exposed Facility of Kibo (Japanese Experiment Module) on the station Sept. 25, the instrument passed several critical onboard tests. The nominal continuous observation of the Earth's atmosphere started Oct. 12. SMILES performed observation for about six months, until a component of the instrument broke down on Apr. 21, 2010.

SMILES detected weak radiation from atmospheric constituents in the submillimeter wavelength region (0.46–0.48 millimeters of wavelength, which corresponds with 624–650 gigahertz of frequency). By calculating the strength of spectral lines coming from various molecular species, the abundance of the species can be retrieved.

One sample (snapshot) of the global distribution of atmospheric constituents retrieved from SMILES observation is the ozone depletion and related change in the amount of chlorine compounds around the North Pole. In winter 2010, planetary waves in the stratosphere became vigorous, resulting in a stratospheric sudden warming event at high latitudes [about 104 °F (40 °C) increase of temperature within several days]. A heterogeneous chemistry process has been underway during this term, and gaseous hydrochloric acid, or HCL, decreased within the polar vortex. At the same time, chlorine chemistry was activated by solar radiation outside the polar night region, resulting in the high chlorine monoxide, or ClO, region. In this region, ozone destruction occurred. SMILES can determine such phenomena with high resolution simultaneously in different regions.

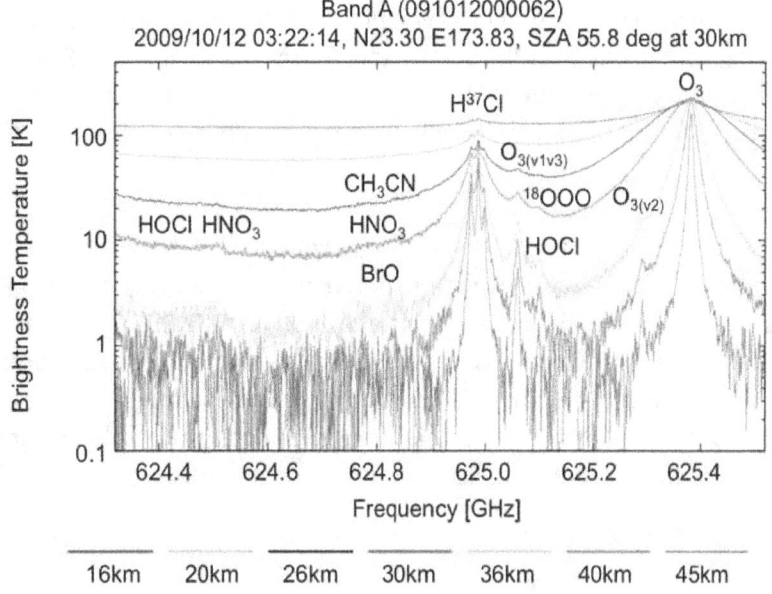

Band A (091012000062)
2009/10/12 03:22:14, N23.30 E173.83, SZA 55.8 deg at 30km

16km   20km   26km   30km   36km   40km   45km

An example of a spectrum observed with SMILES. For comparison of spectra on different tangent heights, the baseline of each spectrum is vertically displaced. Random noise of the spectra is about 0.4–0.6 Kelvin among all tangent heights. (Courtesy of Fujitsu FIP)

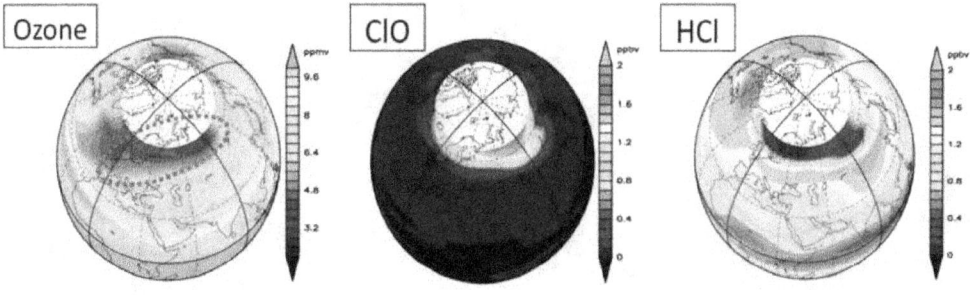

Ozone destruction was activated with ClO and HCl on Jan. 23, 2010. It is shown as the high ClO region (see "ClO" panel). In the high ClO region, O_3 destruction has occurred (see "Ozone" panel). The area where HCl concentration is low suggests the region of polar vortex (see "HCl" panel). (Courtesy of Kyoto University)

In the brightness temperature spectra obtained from SMILES observation data, strong ozone and hydrochloric acid lines with other minor peaks of hypochlorous acid, or HOCl; nitric acid, or HNO_3; bromine oxide, or BrO; and ozone isotopes can be seen. The noise level of the brightness temperature is around 0.4 Kelvin, which is at least one digit smaller than other space-borne limb sounders.

During the first stage of SMILES instrument development, the current literature-based spectroscopic values were used to set the parameters for the spectrometer. The extremely high performance of the spectroscopy with SMILES has led researchers to reevaluate known atmospheric science data and to develop additional atmospheric investigations to test new parameters.

SMILES was successfully launched and performed atmospheric observation with its high spectral resolution. Unfortunately, its operation was terminated after six months of observation; but the observation data indicated the outstanding performance of the SMILES instrument. Preliminary scientific results are reasonable and consistent with the results of other space-borne measurements. It is expected that undiscovered information will be found in the data collected over the six months of operation. Further studies related to data processing and analysis will be necessary to determine the real performance of SMILES instrument.

Using the Space Station To Support Studies Relevant to Understanding Climate Change

Jason Hatton
Science and Applications Division
International Space Station Utilization and Astronaut Support Department
Directorate of Human Spaceflight and Operations
ESA/ESTEC

For many astronauts, the most memorable experience during space flight is seeing the fragile blue globe of Earth beneath the space station. With the changing illumination conditions and passing seasons, one never tires of the dynamic view. Dust storms sweeping the U.S. Southwest, low-pressure areas bringing rain to northern Europe, a typhoon hitting Japan and noctilucent clouds at high latitudes—all these can be easily seen from the orbital outpost. Scientific instruments onboard an international fleet of satellites are routinely sounding, measuring and analyzing the Earth's environment, providing key data for understanding long-term changes in the Earth's climate. To supplement the work of the dedicated Earth observation satellites, the European Space Agency, or ESA, has launched an announcement of opportunity, or AO, for new station experiments for climate change relevant studies.

Various natural physical processes modify the atmosphere, oceans and land surfaces on short and long-term scales. In the past 150 years, human activities have resulted in significant changes in many aspects of Earth's environment, including increases in greenhouse gas concentrations, modification of the nitrogen and phosphorous cycle and major alterations of land use (e.g., deforestation). It is crucial that we understand the interaction of human-caused alterations and natural changes to predict future changes in the Earth's environment. In turn, this information will assist sustainable development in relation to human activities while minimizing degradation of the environment and limiting the vulnerability of society to climate change.

ESA, along with other international agencies, is currently operating a number of Earth observation satellites carrying dedicated instruments to address specific mission objectives. These are supported mainly by ESA's Living Planet Program, the Global Monitoring for Environment and Security, or GMES, Program (jointly carried out with the European Union) and ESA's Climate Change Initiative.

A large variety of international research activities are being performed routinely on board the International Space Station. Historically, the main focus of European research has been in the area of life and physical sciences, taking advantage of the microgravity and exposure to the space environment provided by the station. However, the station has a clear potential to be used as a multiuser platform for studies in astrophysics, solar science, fundamental physics, Earth science and climate change.

Earth framing the International Space Station in May 2010 following undocking of Atlantis during the STS-132 mission. (Image: NASA)

To assess the level of interest of the European and international research communities in deploying remote-sensing instruments on the space station for global change studies, a call for ideas was issued by ESA's Directorate of Human Spaceflight (now Directorate of Human Spaceflight and Operations) and supported by the Directorate of Earth Observation Programs in Oct. 2009. Forty-five proposed ideas were received, with many promising concepts proposed. This confirmed a high level of interest in the use of the station for climate change studies, and several interesting thematic areas were identified. The recent announcement of opportunity proposals will be peer reviewed, and several candidate experiments will be selected for further detailed study and developed for flight on the space station. Further details are available at http://www.esa.int/SPECIALS/HSF_Research/SEMPM17TLPG_0.html.

The space station offers possibilities to fly instruments and experiments without the development of a dedicated satellite platform. The orbit inclination of 51.6° and altitude of 220-250 miles (350–400 kilometers) are different from those of most Earth observation satellites. Instruments can be mounted on a variety of external locations, including the truss structure and dedicated platforms on the European Columbus, Japanese Kibo and Russian segment modules. The European Columbus module's External Payloads Facility, or CEPF, has four payload attachment sites on the end of the module, permitting nadir, zenith and side (limb) viewing. Several instruments relevant to Earth science and climate change are either in development or already deployed on the station. The zenith port of the CEPF is currently occupied by the ESA Sun Monitoring on the External Payload Facility of Columbus, or SOLAR, instrument, which measures the sun's energy irradiance—an important parameter for climate studies. The ESA Atmosphere Space Interactions Monitor, or

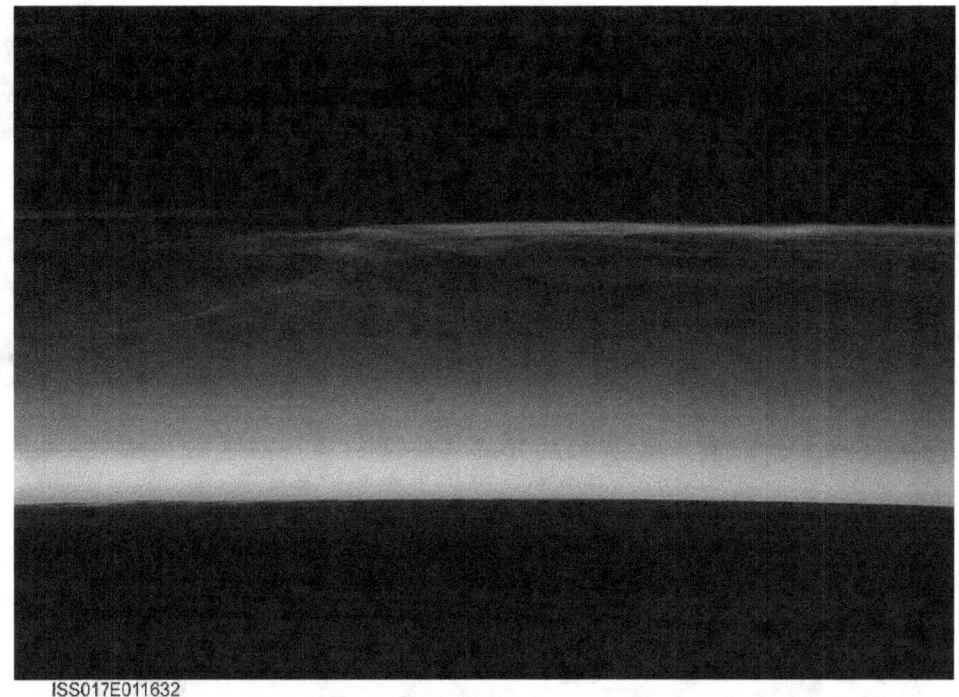

ISS017E011632

Noctilucent clouds photographed from the space station in July 2008. (Image: NASA)

ASIM, which will study high-energy optical and gamma ray events associated with thunderstorms, will be deployed on Columbus in 2015. On the external platform of the Japanese Kibo module, the JAXA Superconducting Submillimeter-Wave Limb Sounder, or SMILES instrument, measured trace gases in the stratosphere, including chemicals that interact with the Earth's ozone layer. Also mounted on the outside of the Kibo, the NASA Hyperspectral Imager for the Coastal Ocean (HICO) is an imaging spectrometer for studying coastal waters.

An additional location for instruments is inside the station, taking advantage of high-quality viewing ports and windows. The NASA Destiny Laboratory has the Window Observational Research Facility, or WORF, which allows viewing through a dedicated nadir-viewing window, while the Cupola module has seven windows providing panoramic nadir and limb views of the Earth. Earth observation imaging using handheld digital cameras is currently performed by the crew members through the station windows as part of the Crew Earth Observations, or CEO, experiment. Potentially, laboratory or airborne instruments could be developed and flown within relatively short lead times. The station provides a normal "shirt sleeve" environment in which to operate instruments as well as the possibility for the crew to interact directly with the experiment (e.g., to change configuration or filters).

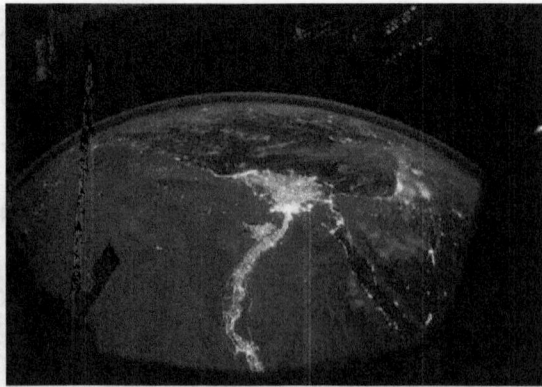

(Right) The Cupola on the International Space Station provides a panoramic view of the Earth for observations. (Left) A night-time image of the Nile Delta region and Eastern Mediterranean taken from the station in November 2010. (Images: NASA)

Investigation of Earth Catastrophes From the International Space Station: Uragan Program

Igor V. Sorokin
S.P. Korolev Rocket and Space Corporation Energia

The Uragan program aboard the Russian segment of the International Space Station uses digital photography to study Earth's natural resources by monitoring catastrophes, both natural and human made. Uragan, which means "hurricane" in Russian, began during the first days of habitation on the station and continues to be an important Earth observation program, with the primary goal of defining requirements for a ground-space system for disaster warning and damage mitigation. The program is a logical continuation of the Earth Visual-Instrumental Observations Program (in other words, a crew Earth observation program) started in the Soviet Union/Russia in the early 1970s as part of the Salyut series of space stations and followed by the Mir orbiting complex.

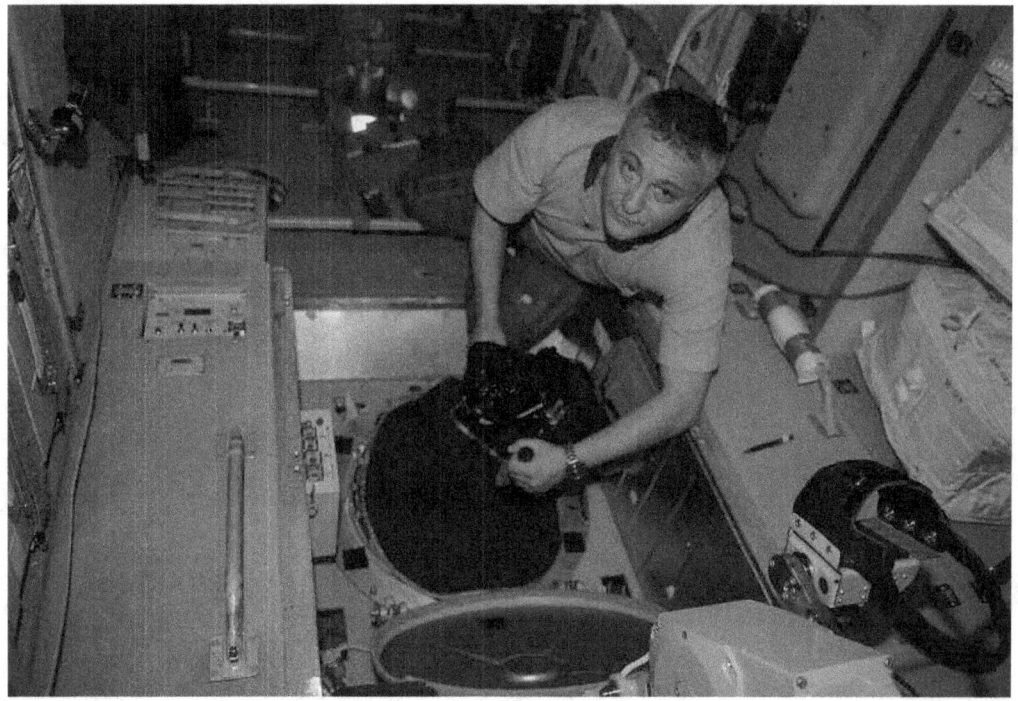

Cosmonaut Fyodor Yurchikhin (Expedition 15) works with the Photospectrometric System or FSS used for Uragan.

In recent decades, humankind has faced various natural and human-made disasters—some with widespread damage. The Uragan program examines natural and human-made catastrophes, including earthquakes, volcanic eruptions, floods, fires, hurricanes, piping accidents and aviation accidents. Information on these types of catastrophes is beneficial to experts in various fields, governments and scientists developing models of catastrophic phenomena and many more. The space station is a convenient platform for Earth observation in that it provides a space for testing a variety of equipment, software and observation methods. Eventually, some of the items tested may be used on robotic spacecraft for remote observations.

Practical needs and quick response are used as the basis for the Uragan program. Images taken from the station can be used by government agencies, scientists, and others to determine the effects of natural or human-made disasters. This program has photographed several disasters since its inception, including human-made oil pollution. In the Caspian Sea region, images taken by station crew members show how oil pollution has affected the coastal areas. Through analysis of the images, three major areas or sources have been identified: the northeast coast in Kazakhstan, southeastern Turkmenistan, and the Absheron region of Azerbaijan. On the northeast coast of the Caspian Sea in Kazakhstan, the large oil deposits formed water-oil lakes ranging in size from several yards or meters up to 7.5 miles (12 kilometers) across. Portions of these lakes are surrounded by ground walls to prevent flooding and contamination of the surrounding areas. The Caspian Sea level continuously fluctuates, and waves up to 9 feet (3 meters) high can be caused by strong winds. Because of the probability of floods on the coast, the potential for failure of these lake walls and corresponding oil flow into Caspian Sea waters is rather high.

Oil pollution in the northern part of the Caspian Sea (on the basis of data received from the Uragan experiment): 40 oilfields, equaling approximately 10 percent of the surface covered with oil film.

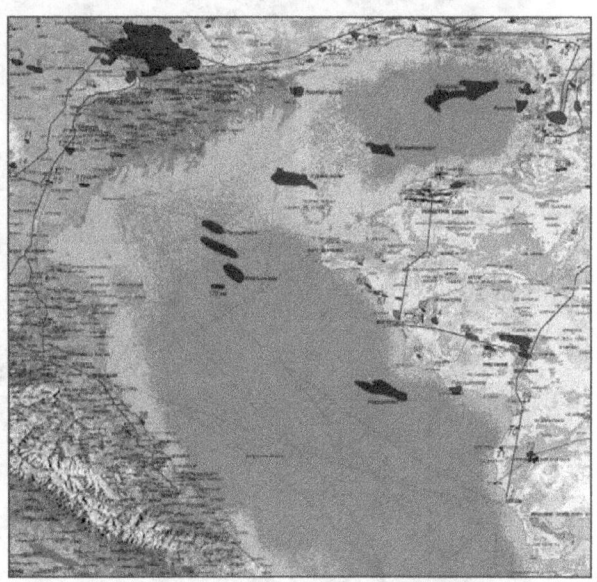

Besides human-made disasters, natural disasters are also monitored. In 2002, crew photography documented important information about a glacier-related disaster in the Caucasus Mountains. On Sept. 20, a small glacier called Kolka in North Ossetia, Russia, unexpectedly released a great amount of ice mixed with water and stones. This mass quickly traveled down the valley causing widespread destruction, loss of life and blocking a hollow in front of Rocky Ridge.

A cosmonaut's photograph taken one month earlier on Aug. 13, 2002, showed the north part of Kazbek Mountain with the small Kolka glacier clearly in view in the upper part of the Genaldon River valley. The glacier was completely covered by stones, giving it a dark appearance. Upper reaches of the Genaldon River before and after the catastrophe were observed using the cosmonaut photographs. This glacial event was unusual, as almost the entire glacier quickly left its bed and streamed down the valley. Scientists around the world observed this rare occurrence. Crew photography taken after the disaster showed a deep, empty hollow that had been full of ice.

The mass media published many differing opinions about the catastrophe, however only the precise detailed surveys from space allowed for a thorough analysis and final determination of the reasons for the disaster. Analysis of the images by experts from the Institute of Geography, Russian Academy of Sciences, discovered clear signs of initial phase of avalanche masses motion. The specific color of ice-soil mixture and changes of structure of the glacier cover indicated melting of the inner layers of the glacier which gives lubrication for the avalanche flume. Such well planned monitoring improves the ability to predict such disasters in advance.

Kolka glacier and avalanche phenomena in the Caucasus.
(Image by space station Expedition 5 crew member Valery Korzun.)

Other important results obtained within the framework of the Uragan program include observations of floods, forest fires, glacial hazards, reservoirs, seaports, icebergs, and other objects and phenomena. Further development of the program is related to the use of additional equipment for observations in the microwave, infrared and ultraviolet ranges, together with development of mathematical models for investigation of catastrophic phenomena. This program continues to capture images of disasters and through analysis, benefiting all humankind.

Principal Investigators:

Mikhail Yu. Beliayev
S.P. Korolev Rocket and Space Corporation Energia

Lev V. Desinov
Institute of Geography, the Russian Academy of Sciences

Space Station Keeps Watch on World's Sea Traffic
European Space Agency

As the International Space Station circles Earth, it has been tracking individual ships crossing the seas beneath. An investigation hosted by the European Space Agency (ESA) in its Columbus module has been testing the viability of monitoring global maritime traffic from the station's orbit hundreds of miles (kilometers) above since June 2010.

The International Space Station as seen from the departing *Atlantis*
space shuttle, May 23, 2010.
(Credit: NASA image S132E012208.)

The ship-detection system being tested is based on the Automatic Identification System, or AIS, the marine equivalent of the air traffic control system.

All international vessels, cargo ships above certain weights and passenger carriers of all sizes must carry "Class A" AIS transponders, broadcasting continually updated data, such as identity, position, course, speed, ship particulars, cargo and voyage information to and from other vessels and shore.

AIS allows port authorities and coast guards to track seagoing traffic, but the system relies on VHF radio signals with a horizontal range of just 40 nautical miles (74 km). This makes it useful within coastal zones and on a ship-to-ship basis, but not in the open ocean; ocean traffic was largely untracked. However, AIS signals travel much further vertically, making the space station an ideal location for space-based AIS signal reception and, therefore, providing the capability of tracking global maritime traffic from space.

Astronauts were instrumental in enabling the COLAIS experiment, which is an in-orbit demonstration project of ESA's General Support Technology Program. Columbus was not originally outfitted with VHF antennas to capture the AIS signals; they were installed on the outside of the module during a spacewalk in November 2009, with the remaining piece of hardware, the ERNO-Box control computer, installed inside Columbus along with the NORAIS receiver in May 2010. The ERNO-Box is itself an orbital demonstration of a new class of space computer developed by Astrium Gmbh, Germany. Astrium was responsible for overall system integration, and contributed the ERNO-Box and a grappling adaptor, or GATOR, used to attach the AIS antenna to Columbus. Antennas were built by AMSAT.

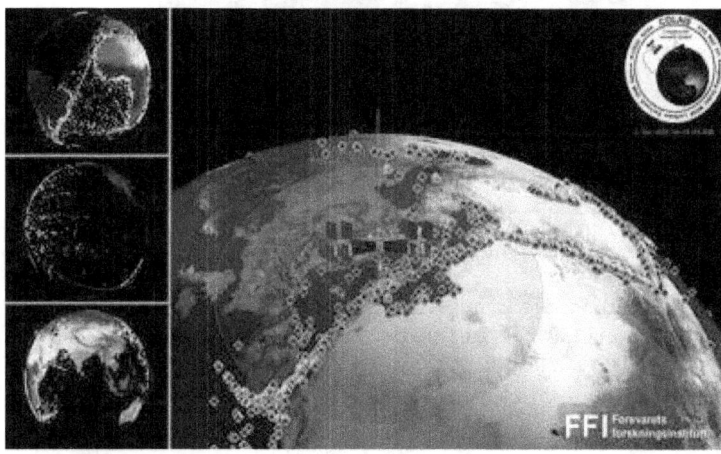

The International Space Station makes 15 orbits around the Earth each day. The illustration of the COLAIS system in operation using the NORAIS receiver shows the station passing the Mediterranean. Its field of view is shown in red; detected vessels are shown as ship symbols, and terrestrial AIS base stations are shown as buildings. The data are for the first five hours of June 3, 2010. (Credit: FFI)

Global Overview of Maritime Traffic

The AIS ground coverage from the station is between approximately 68° north and 68° south. The system consists of two antenna assemblies that were mounted on the outside of Columbus during a spacewalk in November 2009 as well as data relay hardware (the ERNO-Box) and a receiver mounted inside Columbus. The two operational phases with the first receiver from Norway, or NORAIS, which is operated by FFI/Norway, have been extremely successful, with data telemetry received by the Norwegian User Support and Operation Center, or N-USOC, in Trondheim, Norway, via ESA's Columbus Control Center in Germany. Data has been received by NORAIS in almost continuous operation, and all modes of operation have worked extremely well. The NORAIS Receiver has a sample mode that can collect the raw signal, digitize it and send it to ground for analysis of signal quality, which is proving very helpful in making additional improvements/refinements to the system in extremely crowded shipping areas where the possibility of lost signals or mixed signals can occur.

Astronaut Randolph Bresnik seen during Atlantis EVA-2 on November 21, 2009, with the unfurled AIS antenna, attached to Columbus for use in experimental tracking of VHF signals of ships at sea. (Credits: NASA)

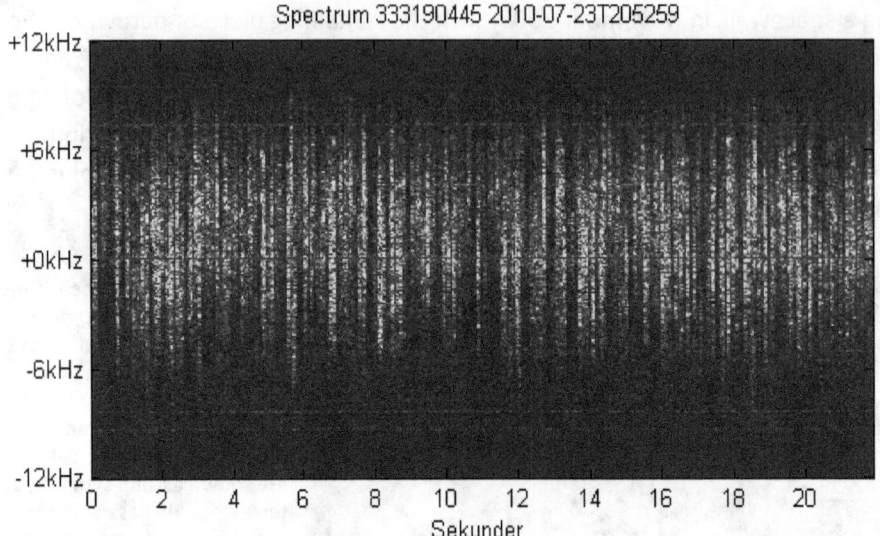

Spectrum 333190445 2010-07-23T205259

The spectrum versus time for 22 seconds of a sampled data. The messages can be seen as vertical lines. Approximately 150 messages are decoded from the data. (Credit: FFI)

This is used both to investigate the signal environment and to evaluate the performance of new receiver technologies on the ground. Several hundred data sets have been collected and processed with new candidate algorithms for next generation receivers.

The results of the analyses have been very good. On a good day, approximately 400,000 ship position reports are received from more than 22,000 different ship identification numbers (Maritime Mobile Service Identity, or MMSI). In a summary made in Oct. 2011, the total number of position reports received exceeded 110 million messages from more than 82,000 different MMSI numbers.

As an addition to the original technical topics, operational experimentation has been included in the investigations. Near-real-time data transfer is crucial to meet the requirement of SAT-AIS set by ESA in cooperation with operational users. After an upgrade of the ground systems in the N-USOC, 10 days of near-real-time data show that 80 percent of the messages collected in the period could be delivered through the station's communications network with data latency significantly less than 1 hour. The near-real-time data delivery has been part of routine operations since Nov. 2011.

At present, a new version of the decoder algorithm, developed by Kongsberg Seatex as part of the technology development contract with ESA, is being tested. The development benefits from the investigations of the sampled data and ongoing work in other ESA projects. The firmware was uploaded to the NORAIS Receiver through the station's communications network and verified and activated in Jan. 2011. The preliminary results indicate that the performance in terms of decoded messages has increased by a factor of between 1.5 and 2.0 for the high traffic zones that ESA has specified should be monitored with high performance.

The work on better algorithms continues. A second NORAIS Receiver upgrade is planned in May 2012. The results of the development will support the design and development of a space-based AIS system in general as well as the performance of the AIS receiver on the station.

Integrating AIS information with other satellite data, such as information from remote-sensing satellites, should significantly improve maritime surveillance and boost safety and security at sea. The payload designed for the Norwegian AISSat-1 satellite, which launched into a near polar orbit in July 2010, provides similarly good data in the high north. The NORAIS Receiver is software-defined radio design operating across the maritime band from 156 to 163 megahertz. The tuning of the NORAIS receiver to frequencies under consideration for allocation to space-based AIS has been carried out, and NORAIS took part in international tests of these two proposed frequencies in October 2010 as arranged by U.S .Coast Guard.

The main reason for covering more than the two current frequencies in use for AIS is to have the possibility to demonstrate the operational use of new channels in the maritime band being allocated to space-based AIS. Also, this configuration allows for characterization of the maritime VHF spectrum with respect to occupancy and interference. The software implementation allows for optimization of the receiver settings in orbit and also allows for upload of new signal processing algorithms.

The Vessel Identification System, or VIS, could potentially be beneficial to many European entities, particularly in assisting them in law enforcement, fishery control campaigns, maritime border control, maritime safety and security issues, including marine pollution surveys, search and rescue and anti-piracy. Various service entities have already been asking to get access to the VIS data, which is continuously acquired on Columbus.

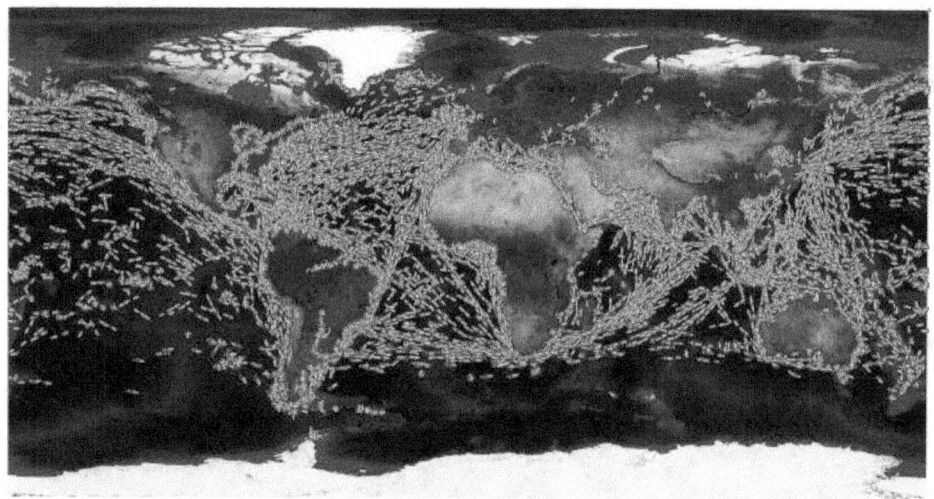

Ship position reports received with the NORAIS Receiver during 24 hours, 29th June 2010.
(Credit: FFI)

Inspiring Youth With a Call to the International Space Station

Jessica Nimon and Camille Alleyne
International Space Station Program Science Office
NASA Johnson Space Center

Ever since the Amateur Radio on the International Space Station, or ARISS, hardware was first launched aboard space shuttle Atlantis on STS-106 and transferred to the space station for use by its first crew, it has been used regularly to perform school contacts. With the help of amateur radio clubs and ham radio operators, astronauts and cosmonauts aboard the station have been speaking directly with large groups of people, showing teachers, students, parents and communities how amateur radio energizes students about science, technology and learning. The overall goal of ARISS is to get students interested in mathematics and science by allowing them to talk directly with the crews living and working aboard the station.

The ARISS conversations usually last for about 10 minutes. During that time, chosen students on the ground ask a preselected set of questions, which the crew answers from aboard the station.

In preparation for these exchanges, students learn about the space station as well as about radio waves and how amateur radio works. Ken Ransom, project coordinator with the space station Ham Radio Program, points out the educational benefits of these communications, of which about 50 a year take place. "The ARISS program is all about inspiring and encouraging by reaching the community and providing a chance for schools to interact with local technical experts. It also brings the space program to their front door."

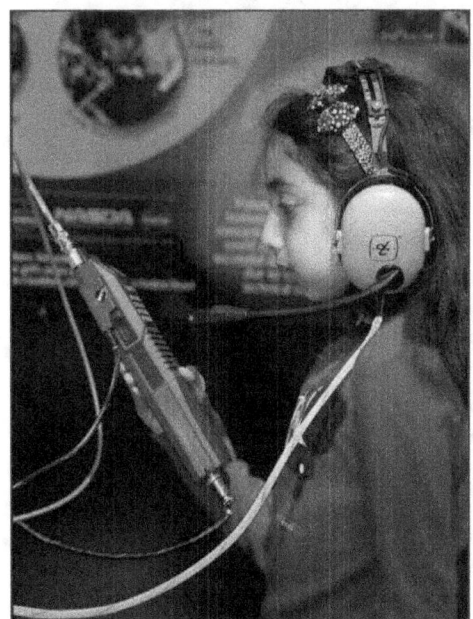

In order for ARISS to work, the station must pass over the Earth-bound communicators during amateur radio transmissions to relay signals between the station's ham radio and ground receivers. Other issues, such as weather and crew availability, also factor into the timing of the transmissions. During this pass, an average of 18 questions can be answered, depending on the complexity of the query. To date, the space station has held more than 600 ARISS sessions with students around the world.

A student talks to a crewmember onboard the International Space Station during an ARISS contact. (Image courtesy of ARISS)

The downlink audio from ARISS talks can be heard by anyone in range with basic receiving equipment; transmissions broadcast on 145.800 megahertz. Interested parties can also catch a broadcast via EchoLink and the Internet Radio Linking Project, or IRLP, amateur radio networks or on the Internet, when available, according to Ransom.

For students who have never thought about the exploration of space, being involved in an amateur radio event such as this can be an eye-opener and pave the way for them to dare to dream and for those dreams to come true.

U.S. educators interested in participating in an ARISS communication can contact NASA's Teaching From Space Office for a proposal packet. International schools should submit applications via the ARISS Web site for consideration. Submissions are due in July and January of each year.

Astronaut Sunita L. Williams, flight engineer for Expeditions 14 and 15, talks with students at the International School of Brussels in Belgium during an Amateur Radio on the International Space Station (ARISS) session in the Zvezda Service Module.
(NASA image ISS014E18307)

Students Get Fit the Astronaut Way

Jessica Nimon, International Space Station Program Science Office
NASA Johnson Space Center

When you think of NASA, you likely picture the space shuttle or International Space Station or have images of planets and galaxies flashing before your mind's eye. NASA's Mission X: Train Like an Astronaut, however, focuses a little closer to home. Working with the schools in our very own neighborhoods and around the world, Mission X uses the same skills used to train astronauts to motivate physical education for around 3,700 students in 40 cities around the globe.

The brainchild of the International Space Life Science Working Group, or ISLSWG, and the Human Research Program Education and Outreach, or HRPEO, Mission X launched in U.S. schools Jan. 18, 2010. NASA's Human Research Program funded the pilot program, including activity and educational modules and an interactive Web site (www.trainlikeanastronaut.org). The program is available in six different languages for participants in 10 countries—the U.S., the Netherlands, Italy,

Students at Daltonbasisschool de Tjalk in Lelystad, Netherlands, participate in the building an astronaut core activity. (Courtesy of ESA)

France, Germany, Japan, Australia, Columbia, Spain, and the United Kingdom. The goal of the program is to make kinesiology and nutrition fun for children by encouraging them to train like an astronaut.

Chuck Lloyd, the NASA program manager responsible for the project, comments on how the space program excites students, prompting active participation. "Mission X is all about inspiring and educating our youth about living a healthy lifestyle with a focus on improving their overall daily physical activity with the Mission X physical activities, known as *train-like-an-astronaut*."

Students ranging from 8 to 12 years old learn about the science behind their activities, including the importance of hydration, bone health, and balanced nutrition. Known as "fit explorers," these youth stay motivated with fun ways to gauge their success. For instance, they can see what other schools are doing on the Train Like An Astronaut blog. Fit explorers logged their accumulated activity points over the course of the program to help an online cartoon astronaut, known as Flat Charlie, Walk to the Moon. Flat Charlie made it to the Moon five weeks into the competition—a distance of 238,857 miles (384,403 km) or 478 million steps—and he's still going!

Fit explorers learn that astronauts train before, during and after missions to maintain top physical health via good nutrition, rest and physical activity habits to function in the demanding environment of microgravity. Lloyd makes the connection of such health-centric mindsets for everyone, even those not planning to launch into space. "Our youth must also make smart choices on balancing the amount of work, play and sleep they get to remain in peak performance. Education is critical to our youth and to our communities to ensure we have tomorrow's workforce and technical leadership to address the rigors of our societies."

The first phase of the challenge, which lasted for six weeks, had schools participating with each other within their own countries in a friendly competition. The second phase of the challenge will be an ongoing multi-year Mission X Train Like An Astronaut challenge that will be expanded to more students and more countries.

Students at Media Sandro Pertini school in Savona, Italy, participate in the Building an Astronaut Core activity. (Courtesy of ASI)

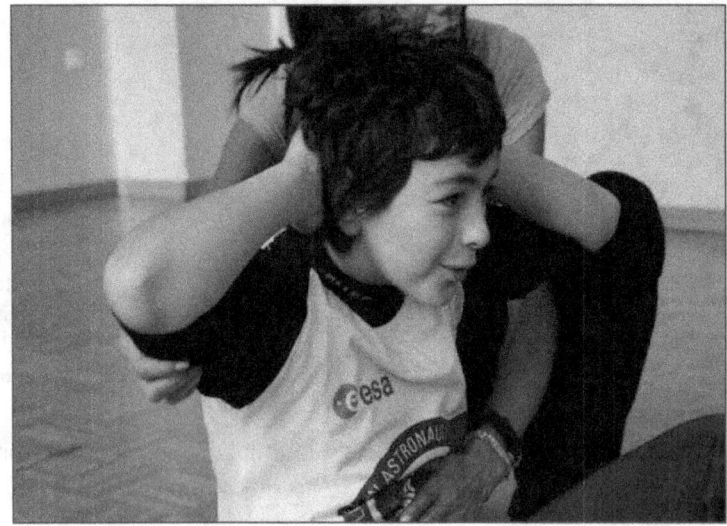

Europe's Alliance With Space Droids

Nigel Savage, Ph.D., Head of Education Unit
Directorate of Human Spaceflight and Operations
European Space Agency

Between video games and sci-fi movies, a robotic squadron of miniature satellites come to life aboard the International Space Station, obeying the commands of young students.

The European Space Agency (ESA) is participating in the NASA and Massachusetts Institute of Technology, or MIT, Zero-Robotics competition, a chance for high school students to program droids for action on the space station. SPHERES—also known as Synchronized Position Hold, Engage, Reorient, Experimental Satellites—are volleyball-sized satellites with their own power, propulsion, computers and navigation.

The challenge to remotely control them began in the United States, where an adventurous professor from MIT found inspiration in the Star Wars saga to create these intriguing robots. The mini-spacecraft have been used inside the station since 2006 to test autonomous rendezvous and docking maneuvers.

Now formation flying in zero-gravity becomes a programming issue for European students, also. A number of schools from ESA member states create rival programs that control three SPHERES in real time on the space station.

A local SPHERES expert, familiar with the coding requirements for the droids, is assigned to each European school. Sponsored by ESA, several university staff members are being trained at MIT.

The competition is not only about feeding the satellites sets of commands. The local experts help students build critical engineering skills, such as problem solving, design thought process, operations training and teamwork. Their results could lead to important advances for satellite servicing and vehicle assembly in orbit.

Teams in the U.S. and Europe test their algorithms under realistic microgravity conditions by competing in elimination rounds against each other with finals on both sides of the Atlantic.

The winners' software is uploaded and run in the three weightless SPHERES by astronauts on the station. The exciting final is streamed live at ESA's technology center in the Netherlands, known as ESTEC, and MIT.

This is the start of Europe's alliance with the space droids, the first in a string of global education projects with the station as a common scientific platform for students worldwide. If the pilot experience proves successful this year, ESA anticipates including larger numbers of participants from all over Europe in future competitions. And may the force be with them.

View of the Synchronized Position Hold, Engage, Reorient, Experimental Satellites (SPHERES) floating in the Destiny laboratory module as seen by the Expedition 14 crew. Flight Engineer Thomas Reiter is visible in the background.
Image credit: NASA.

Water: A Chemical Solution

Camille Alleyne
International Space Station Program Science Office
NASA Johnson Space Center

Can you imagine not having access to safe and clean drinking water? Well, that is the plight of hundreds of millions of people around the world. One in eight people lacks access to this fundamental resource for human survival.

The most abundant substance on the Earth's surface, water, H_2O, is essential for life. It covers about 70 percent of the planet's surface, and it makes up about 70 percent of the human body. Clean water is essential for human health and well-being.

Space engineers who design the International Space Station systems share the same need as those concerned about the lives of people in remote areas of Earth—to be innovative in developing ways to make this vital resource readily available. Thus, the goal of the 2011 Global Water Experiment is to educate students about the chemistry of water resources around the world and to increase awareness of how people in different environments use various methods to provide clean, safe drinking water.

Rita Nobile, 15, of Pacifica High School, speaks to members of NASA's Digital Learning Network as part of a Webcast. Photo by Anthony Plascencia, Ventura County Star.

The United Nations Educational, Scientific and Cultural Organization or UNESCO, in partnership with the International Union of Pure and Applied Chemistry, or IUPAC, designated 2011 as the International Year of Chemistry. As a part of the year-long activities, students across the world will sample water in their communities and conduct experiments to learn about our planet's acidity levels, water salt content, and various water treatment, disinfection and salt removal method. The results will be used to build an interactive, global water data map.

The American Chemistry Society is sponsoring Water: A Chemical Solution, providing students in the United States with an opportunity to learn not only about the water challenges other people face, but to encourage interest and generate enthusiasm for chemistry among young people and to promote international cooperation toward a creative future for chemistry.

The experience helps students learn how the International Space Station water resources are managed and about the fundamental technology behind the station's Environmental Control Life Support System. The system handles up to 23.2 pounds of condensate, crew urine and urinal flush water to produce a purified distillate. This distillate is combined with other wastewater sources collected from the crew and cabin and is processed, in turn, by a water processor assembly that ultimately produces drinking water for the crew.

The station's water processor has an iodinated resin that is used in the filtration process. Iodine is added to the water to control the growth of microorganisms—in the same way chlorine is added to the water we drink at home. Iodine is used instead of chlorine because iodine is much easier to transport to orbit, and because it is less corrosive. This iodinated resin has been developed as a commercial water filtration solution for use in disaster and humanitarian relief zones in a number of countries throughout the world.

Water recycling is critical to reducing resupply requirements for human exploration in the extreme environment of space. As a result, students around the world who participate in this Global Water Experiment will also benefit from the knowledge of station's system design as they learn to think critically about the natural world around them and about the most fundamental resource necessary for human survival.

Water recycling system used on the International Space Station. Image credit: NASA/Dimitri Gerondidakis

Ucyu Renshi (Space Poem Chain): Connecting Global People With Words

Masato Koyama
Space Environment Utilization Center, JAXA

For as long as we can imagine, the heavens have at times been an easel on which human beings painted their dreams, sometimes mirroring their own lives. Today, as scientific progress helps unravel the world's mysteries one by one, the skies *above*, extending to outer space, continue to inspire us with limitless curiosity as well as awe of the infinite beyond.

Ucyu Renshi (space poem chain) targets the establishment of a collaborative venue through renshi (chain poetry) by collectively considering space (the universe, Earth and life itself), unfettered by barriers of nation, culture, generation, profession, and position or rank.

JAXA: An Uchu Renshi Symposium is held every year after the completion of Ucyu Renshi to publicize them.

The Japan Aerospace Exploration Agency, or JAXA, started the Ucyu Renshi program to connect people, including crew members in space, with words and allow them to feel more closely involved in space activities. Even those not interested in space sciences and technologies can participate in and enjoy Ucyu Renshi.

Renshi (chain poetry), a form developed from traditional Japanese "renga and renku (linked verse)" in the early 1970s, is now well known and has gained a global following. It is open ended, aiming to fuse the traditions of classical Japanese poetic forms with the world of modern poetry. Renshi is compiled by weaving words like batons in a relay race from one participant to the next.

Ucyu Renshi is a form of renshi compiled by thoughts of imaging space, including Earth and our lives. Collaborating with and considering other participants is clearly essential to creating remarkable Ucyu Renshi, and doing so helps to build bonds among people.

JAXA compiles one Ucyu Renshi collection every year, consisting of around 24 short poems. Half are written by the public, while the remainder feature the contributions of famous poets. After the completion of the Ucyu Renshi, it is recorded on a DVD, which is then loaded into the International Space Station every year. A Ucyu Renshi symposium is also held to introduce the Ucyu Renshi to the public. JAXA also is applying Ucyu Renshi to Japanese language classes in elementary schools. Students compile Uchu Renshi in class and learn how to create the poems with famous poets while learning the importance of cooperation within the class. Students enjoy creating Ucyu Renshi and are very excited about participating in the "Kibo" big space project.

JAXA: Ucyu Renshi URL: http://iss.jaxa.jp/utiliz/renshi/

JAXA started the Ucyu Renshi in 2006, and the four Ucyu Renshi already composed are now stored in the International Space Station's Japanese Experiment Module, or "Kibo," which means "hope" in Japanese. All participants can watch the space station from the ground, imagining their poems stowed in the station, a shining star in space. Since it started in 2006, the number of participants has steadily increased, with more and more people interested in Ucyu Renshi and space activities. Furthermore, Ucyu Renshi inspires them with a new universal view.

Students Photograph Earth From Space via EarthKAM Program

Arun Joshi, International Space Station Program Science Office
NASA Johnson Space Center

Imagine this: you are a young and inquisitive middle-school student, investigating and examining the vastness of Earth's majestic mountain ranges, coastlines, oceans and other geographic imprints. Now, envision the thrill of doing so from the vantage point of an astronaut! Earth Knowledge Acquired by Middle School Students, or EarthKAM, allows students to do just that—view and capture images of their world from an astronaut's perspective!

EarthKAM is a NASA-funded educational outreach program run in collaboration with the University of California at San Diego. The goal is to provide an enriched and enhanced educational experience to motivate students toward math and science studies. The camera allows students worldwide to examine and photograph Earth from the unique vantage point of the International Space Station.

EarthKAM uses a Nikon D2Xs digital camera mounted in the Window Observational Research Facility, or WORF, which uses the science window located in the U.S. Destiny Laboratory. This window's high quality optics capabilities allow the camera to take high-resolution photographs of the Earth using commands sent from the students via the online program. Students and educators then use the photos as supplements to standard course materials, offering them an opportunity to participate in space missions and various investigative projects. Creators of EarthKAM hope that

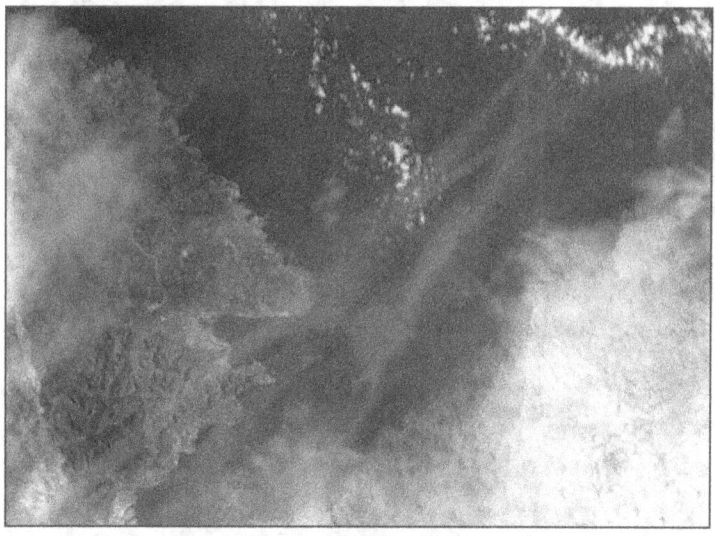

The above photo of Croatia is one of the final pictures taken during the most recent EarthKAM session in July 2011.
(Image courtesy of EarthKAM)

combining the excitement of this space station experience with middle-school education will inspire a new set of explorers, scientists and engineers.

Students use EarthKAM to learn about spacecraft orbits and Earth photography through the active use of Web-based tools and resources. With the help of their teachers, they identify a target location and then must track the orbit of the station, reference maps and atlases and check the weather prior to making their image request. These requests funnel to another set of students, this time at the University of California at San Diego. These college students run the EarthKAM Mission Operations Center, or MOC, for the project. Here they compile the requests into a camera control file and, with the help of NASA's Johnson Space Center, then uplink the requests to a computer aboard the space station.

Requests ultimately transmit to the digital camera, which then takes the desired images and transfers them back to the station computer for downlink to EarthKAM computers on the ground. This entire relay process usually completes within a few hours, and the photos are available online for both the participating schools and the public to enjoy.

As an added bonus, EarthKAM does not require much attention aboard the space station, which allows the astronauts to pay more attention to the other more involved payloads. According to Annie Powers, a NASA flight controller in the Cargo Integration and Operations Branch at the Johnson Space Center, "The crew's main role is the set up: they position the camera on a bracket over the window, adjust the camera settings, connect the USB cable to the laptop and start the EarthKAM software. But, after set-up, all the crew has to do is periodically change the camera battery, and we usually have them swap to a different lens mid-week. It's a very autonomous system; it pretty much just snaps away!"

The above photo of North Korea is from a series of images that UCSD
students annotate for EarthKAM participants. The photo was taken during
the winter 2011 session of EarthKAM. (Image courtesy of EarthKAM)

In its earliest stages, the EarthKAM program went by the name of KidSat, coined by NASA astronaut and creator of the program Sally Ride. EarthKAM flew on five shuttle flights prior to its relocation to the space station in 2001. The EarthKAM camera has since been a permanent payload aboard the space station and supports approximately four missions annually. July 2011 marked the most recent session of EarthKAM.

The EarthKAM program brings education out of the textbooks and into real life for students. From its first space station expedition in March 2001 to now, EarthKAM has touched the lives of over 165,000 student participants and an unknown number of online followers. The program also has a strong international presence with users from 41 countries to date. "With every new mission," added Powers, "you can just see the numbers growing because [EarthKAM] is so automated now!" Interested viewers of EarthKAM images and educators interested in participating have the opportunity to register online.

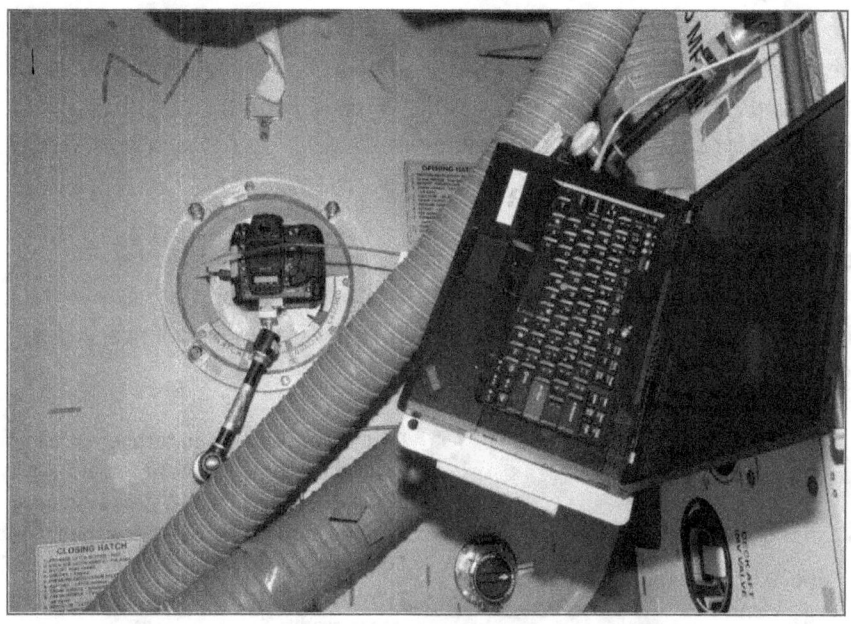

The EarthKAM camera mounted on a multiuse bracket with a clamp at the Node 2 nadir hatch window taken during the EarthKAM session in July 2011. (ISS028E019379 courtesy of NASA)

Red Food for the Red Planet

Canadian Space Agency

Among the many issues that space programs face as they develop plans to send a human mission to Mars, the question of life support ranks at or near the top. It should come as no surprise, then, that this challenge is at the heart of research being conducted by some of the world's top scientists, including those in Canada.

These scientists are investigating options that would provide astronauts with a self-contained biological system required to sustain life during a long-duration mission. In the case of Mars, planners need to consider a minimum of a six-month journey in each direction to get to and from the planet, coupled with a stay likely somewhere in the neighborhood of eighteen months!

One of the most obvious options is to use plants, which provide food, water and oxygen, while at the same time recycling carbon dioxide and waste. The answer, although seemingly obvious, requires more in-depth thought. Scientists still must wrestle with the questions of which types of seeds are best suited for space missions and whether the environments of both deep space (for the journey), and Mars itself will affect the ability of the seeds to germinate.

In an effort to get Canadian students to address these issues, as a catalyst for inspired and authentic learning, the Canadian Space Agency (CSA) has been collaborating with the University of Guelph, Agriculture and Agro-Food Canada, the Ontario Centres of Excellence, Heinz Canada and Stokes Seeds on the Tomatosphere Project.

During a previous Tomatosphere program, students studied the growth of their tomato plants in Miss Smith's grade three class at Langley Fundamental Elementary, Vancouver, British Columbia, Canada. The students took their plants home to grow in their gardens over the summer. Image credit: Tomatosphere.

Each year, students are provided with sets of tomato seeds that have been exposed to space or space-simulated environments as well as a control group of seeds. Recently, 600,000 of these seeds were flown onboard the final U.S. space shuttle mission to the International Space Station, where they will remain for up to three years prior to returning to Earth and being distributed to more than 13,000 participating classrooms. The project's baseline experiment investigates the germination rate of the seeds; however, supporting materials have been developed to allow educators from grades 3 to 10 to build on student understanding of a variety of topics, from the science of plants to the science of nutrition to the science of ecosystems.

Marilyn Steinberg, the Space Learning Program Manager at the CSA, commented that, through their participation in Tomatosphere, "These students become space-farmers, conducting experiments in their classrooms that teach them about the complexity of horticulture, inspire innovative thinking about food production off-planet, and build their scientific skills set as they prepare to become Canada's first generation of planetary explorers."

Over the next three years, participating students will continue to provide scientists with data related to the possibility of growing tomatoes in space, while developing a "taste" for science and space research.

Those interested in finding out more about Tomatosphere are encouraged to visit the project website at http://www.tomatosphere.org/.

Calling Cosmonauts From Home!

Sergey Avdeev, Deputy Department Head, Division Chief
The Central Scientific Research Institute of Machine-Building – TSNIIMASH

Educating future generations of scientists, technologists, engineers and mathematicians is a global effort—one that includes the contributions of the Russian Federal Space Agency, or Roscosmos. One of the main objectives of activities aboard the International Space Station is the implementation of educational and outreach projects that contribute to attracting young people to study science. These projects also help create modern high-technology equipment and increase support in society for space programs in general and the space station program in particular. Currently on board the Russian segment of the station are four space investigations that have educational components. Coulomb Crystal, Shadow-Mayak, MAI-75 and Great Start continue to demonstrate great benefits in capturing the imagination of students across the Russian region.

Coulomb Crystal is an investigation aimed at studying the dynamics of solid dispersed environments in an inhomogeneous magnetic field in microgravity. Pilot studies onboard the station explore the structural properties of Coulomb clusters—liquid crystal phase transitions, wave processes and the physical and mechanical characteristics of its heating mechanism, to name a few. Students at all levels of schooling, including secondary school and college, have had the opportunity to prepare and conduct the experiment on the ground.

Shadow-Mayak is a VHF radio beacon that allows amateur radio enthusiasts to communicate with crew on board the station. The presence of this equipment on board the Russian segment of the station serves as a learning tool for students in the area of space communications. They study the conditions of the admission-transfer of the radio beacon using the world amateur radio network. They also study the characteristics and spatial distribution of the intensity of the radio broadcast and rebroadcast from the onboard transceiver transmitter.

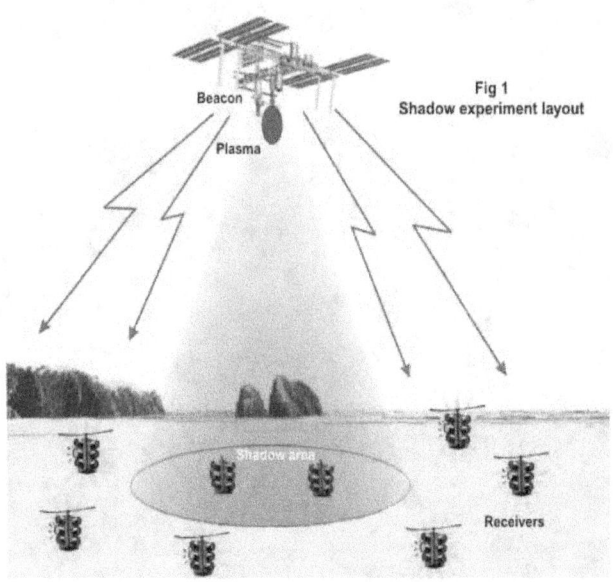

Diagram of the experiment "Shadow-Mayak."
(Image courtesy of ROSCOSMOS)

Along the lines of Shadow-Mayak, MAI-75 also is part of the suite of communication equipment that is housed on board the Russian segment of the station. It allows for a system of quick video downlinks from space in near-real time. This network distribution affords students and amateur radio operators from all over Russia the opportunity to learn first hand from

Image taken by the reception and processing Centre MAI, Courtesy of ROSCOSMOS.

space explorers what it is like to live and work in space. The use of images of Earth from space also is effective in the learning process and serves to inspire and motivate.

Great Start is an investigation aimed at popularizing the achievements of cosmonautics in Russia and in the world. Developed with the preparation of a special questionnaire, this experiment allows the general public an opportunity to express its attitude to the great event in human history—the first human flight in space—as well as to get acquainted with the results of scientific investigations being conducted on board the station. Great Start promotes and enhances international cooperation on the station for further integration of Russia into the world of cultural, educational and scientific relations. As a result, there will be scientific and educational workshops held to popularize the achievements of Russian human spaceflight with the involvement of the general population—including students and specialists in various areas of possible utilization of the results of space missions.

There are several more experiments planned within the framework of the educational program: ecology-education, which include student experiments that demonstrate the conduct of airborne microscopic particle suspensions in microgravity; chemistry-education, which include student experiments that capture microgravity structural elements in the specified form on the basis of polymer composite materials and diffusion; and diffusion, which is an educational demonstration of the process of diffusion in liquid environments in weightlessness. These educational projects involve hundreds, if not thousands, of students from all regions of Russia. Like all of our space station global, regional and national education programs, these serve to inspire and motivate generations to study and consider careers in science, technology, engineering and mathematics.

National Aeronautics and Space Administration
http://www.nasa.gov/iss-science/

Canadian Space Agency
http://www.asc-csa.gc.ca/eng/iss/default.asp

European Space Agency
http://www.esa.int/esaHS/iss.html

Japan Aerospace Exploration Agency
http://iss.jaxa.jp/en/

Roscosmos – Russian Federal Space Agency
http://knts.rsa.ru
http://www.energia.ru/english/index.html